Inventory Parameters

Sanjay Sharma

Inventory Parameters

Springer

Sanjay Sharma
National Institute of Industrial Engineering
(NITIE)
Mumbai, Maharashtra
India

ISBN 978-981-10-4544-8 ISBN 978-981-10-4545-5 (eBook)
DOI 10.1007/978-981-10-4545-5

Library of Congress Control Number: 2017937284

Printed on acid-free paper

This Springer imprint is published by Springer Nature
The registered company is Springer Nature Singapore Pte Ltd.
The registered company address is: 152 Beach Road, #21-01/04 Gateway East, Singapore 189721, Singapore

Preface

Inventory management deals with a wide variety of parameters and their interaction. In order to deal with the management of inventories, there is an important need for a thorough understanding of relevant parameters. The present book specifically focuses on the inventory parameters and that makes it different from the existing books on inventory management. The various parameters considered in this book include the costs of ordering, inventory carrying, and stock out, as well as the procurement of inventory items. Demand and procurement cycle time are also incorporated. Finally, it is expected that this book will prove to be a significant reference/text for students/professionals of management and engineering branches including production/procurement management.

Mumbai, India

Sanjay Sharma

Contents

1 Introduction 1
1.1 Parameters 3
1.1.1 Demand 3
1.1.2 Procurement Cost 5
1.1.3 Cycle Time 6
1.1.4 Ordering Cost 6
1.1.5 Inventory Carrying Cost 8
1.1.6 Inventory Stock 9
1.1.7 Stock Out Level 9
1.1.8 Stock Out Cost 10
1.2 Economic Lot Size 11
1.3 Interaction of Inventory Parameters 20
1.3.1 Basic Case Without Stock Outs 21
1.3.2 With Stock Out 22

2 Demand 25
2.1 Increase in Demand 27
2.2 Decrease in Demand 30
2.3 Stock Out Situation 32
2.3.1 Demand Increase 34
2.3.2 Demand Decrease 36
2.4 Demand Estimation for Variety of Products 40
2.4.1 FMCG 41
2.4.2 Pharmaceutical 41
2.4.3 Agricultural 42
2.4.4 Hi-Tech 43

3 Ordering Cost 45
3.1 Estimation of Ordering Cost 45
3.2 Ordering Cost Decrease 48
3.3 Ordering Cost Increase 51

3.4 Inclusion of Shortages . . . 54
3.4.1 Reduction in Ordering Cost . . . 54
3.4.2 Increase in Ordering Cost. . . . 57
3.5 Interaction of Ordering Cost and Demand . . . 58
3.6 Ordering Cost Estimation for Variety of Products . . . 67
3.6.1 FMCG . . . 67
3.6.2 Pharmaceutical . . . 67
3.6.3 Food Products . . . 68
3.6.4 Hi-Tech . . . 69

4 Inventory Carrying Cost . . . 71
4.1 Carrying Cost Estimation. . . . 71
4.2 Increase in Holding Cost . . . 73
4.3 Decrease in Holding Cost . . . 76
4.4 Backordering Scenario . . . 80
4.4.1 Carrying Cost Increase . . . 80
4.4.2 Carrying Cost Reduction . . . 82
4.5 Interaction of Holding Cost with Other Parameters . . . 83
4.6 Response with a Combination of Parameters . . . 97
4.6.1 Without Stock Outs . . . 97
4.6.2 With Stock Outs. . . . 102
4.7 Carrying Cost Estimation for Various Products . . . 108
4.7.1 FMCG . . . 108
4.7.2 Pharmaceutical . . . 109
4.7.3 Food/Agricultural Products. . . . 109
4.7.4 Hi-Tech . . . 110

5 Procurement Cost. . . . 111
5.1 Procurement Cost Reduction . . . 112
5.2 Procurement Cost Increase. . . . 115
5.3 Stock Out Inclusion . . . 119
5.3.1 Decreased Purchase Cost . . . 120
5.3.2 Increased Purchase Cost. . . . 122
5.4 Interaction of Procurement Cost with Other Parameters. . . . 124

6 Cycle Time . . . 131
6.1 Significance of Cycle Time . . . 131
6.2 Cycle Time Variation. . . . 135
6.2.1 Reduced Cycle Time . . . 137
6.2.2 Increased Cycle Time. . . . 139
6.3 Backlogging Situation . . . 142
6.3.1 Upward Variation in Cycle Time . . . 147
6.3.2 Downward Variation in Cycle Time. . . . 151
6.4 Interaction of Parameters Without Stock Out. . . . 155
6.4.1 Decreased Cycle Time . . . 155
6.4.2 Higher Cycle Time. . . . 157

6.5 Interaction of Parameters with Stock Out 161
6.5.1 Increased Cycle Time 161
6.5.2 Lower Cycle Time 171

7 Stock Out 187
7.1 Stock Out Cost Estimation 187
7.2 Shortage Cost Increase 190
7.3 Shortage Cost Decrease 193
7.4 Interaction with the Parameters 197
7.4.1 Interaction with the Demand 198
7.4.2 Interaction with the Ordering Cost 201
7.4.3 Interaction with the Holding Cost 205
7.5 Variation of K Triggered by Change in Other Parameters 209
7.5.1 Demand Variation 209
7.5.2 Ordering Cost Variation 211
7.5.3 Holding Cost Increase 212
7.6 Response with a Combination of Parameters 214
7.6.1 With an Aim of Similar Lot Size 214
7.6.2 With an Aim of Similar Total Cost 223
7.7 Stock Out Cost Estimation for Various Products 232
7.7.1 FMCG 233
7.7.2 Pharmaceutical 233
7.7.3 Agricultural 233
7.7.4 Hi-tech 234

Index 235

About the Author

Dr. Sanjay Sharma is a Professor at National Institute of Industrial Engineering (NITIE), Mumbai. As an operations and supply chain management educator and researcher, he has three decades of experience including industrial, managerial, teaching/training, consultancy, and research and has a number of awards and honors to his credit. He has published three books and many papers in various journals such as the *European Journal of Operational Research, International Journal of Production Economics, Computers and Operations Research*, and *International Journal of Advanced Manufacturing Technology*. He also serves as a reviewer for several international journals and is on the editorial board of various journals including the *International Journal of Logistics Management*.

Chapter 1
Introduction

There are several customers/consumers purchasing various products from a retailer/shopkeeper. A retailer or shopkeeper purchases several products from the wholesaler or the company producing them. For example, these products might be packaged sugar/tea leaves, soap, bulb, and pen among others. Such items or products may be stored as several inventory items. Inventory management deals with various issues, some of which are as follows:

(i) Purchase of products in suitable quantities
(ii) Stocking those at appropriate places
(iii) While selling the products, review of the stock

Suppose that the retailer purchases the products in larger quantities. There might be a chance of products remaining unused for long time. Number of customers desiring that product in larger quantities is not available. In other words, the demand at that high level does not exist in the current scenario. On the other hand, suppose that the retailer purchases the products in smaller quantities. There might be a chance of product unavailability if customer demands it. In other words, that particular quantity of product is already sold before the visit of this customer at the retailer/shopkeeper site.

In case where larger quantities are purchased, inventories are carried for longer time, and the money of retailer is blocked for longer time because of no profit gain by selling such quantities. Furthermore, there is a chance of deterioration in the product quality in some of the cases. During that long time, expiry date may also be over. In case where smaller quantities are purchased, these may be sold earlier. Some of the customers may not find the product available during their visit, and there is loss of profit for the retailer. There is dissatisfaction among some of the customers and also loss in the potential gain for the retailer. In other words, stock out scenario is faced by all concerned. The effects of purchasing in larger/smaller quantities are represented by Fig. 1.1.

S. Sharma, *Inventory Parameters*, DOI 10.1007/978-981-10-4545-5_1

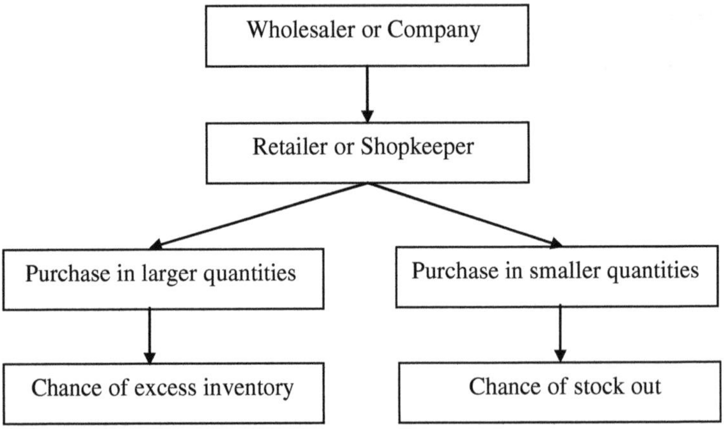

Fig. 1.1 Retailer purchase in larger/smaller quantities

While purchasing the products in suitable quantities (i.e., neither larger nor smaller), the attention should be paid on the following factors among others:

(a) Demand of the customer
(b) Change in the customer demand
(c) Time taken in the replenishment
(d) Efficiencies concerning the source of procurement and delivery

After arrival of the products at the retailer/shopkeeper location, care should be taken in stocking those products. The items for which demand is higher may be kept nearer to the billing location. This helps in convenient frequent access, and also the handling efforts are less. Some of the items may not be kept where sunshine is available. Stocking may depend on various aspects, some of which are as follows:

(i) Nature and characteristics of the product
(ii) Weight of the product
(iii) Size of the item
(iv) Material by which product is made
(v) Shape of the product
(vi) Sales volume
(vii) Packaging of the product

During the sales of an item, review of the stock is necessary. As the sales take place, stock of an item reduces. For example, as shown in Fig. 1.2, current stock is 20 products. As the sales progress, stock reduces to 14 since sale of six products happens on Day 1. At the end of the Day 2, stock reduces to seven because of the sale of seven products.

Observing the sale of 6–7 products, and assuming that the products will arrive in one day time after placing an order for them, the retailer/shopkeeper should inform

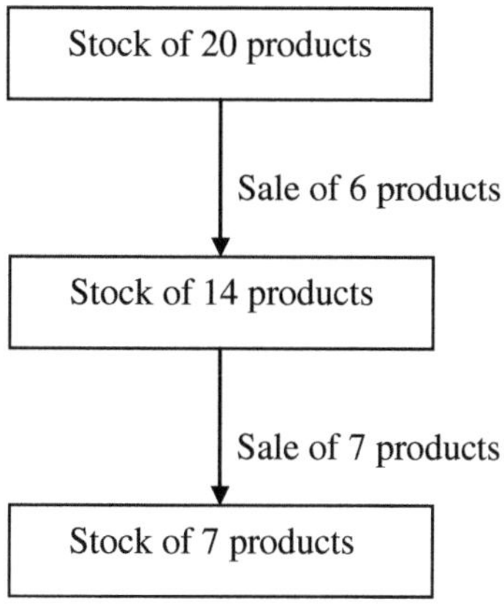

Fig. 1.2 Stock reduction with sales progress

the company/wholesaler to send the required quantities of that item now. In other words, an order should be placed now when the stock reduces to seven products.

For an effective inventory management, it is very necessary to understand the different types of factors or parameters involved in the process of analysis.

1.1 Parameters

Various types of factors or parameters are shown in Fig. 1.3.

1.1.1 Demand

Suppose that the ten kg of sugar is consumed in a house every month. Therefore, the monthly demand of sugar of that house is ten kg. For estimating the demand of such item concerning a shopkeeper, assume that ten such house representatives visit the shop every day. Therefore, an approximate:

(a) Daily demand = $10 \times 10 = 100$ kg
(b) Weekly demand = $100 \times 7 = 700$ kg
(c) Monthly demand = $100 \times 30 = 3000$ kg
(d) Yearly demand = $100 \times 365 = 36500$ kg

The above mentioned approximate demands can also be written as:

(i) Demand per day
(ii) Demand per week
(iii) Demand per month
(iv) Demand per year, or an annual demand

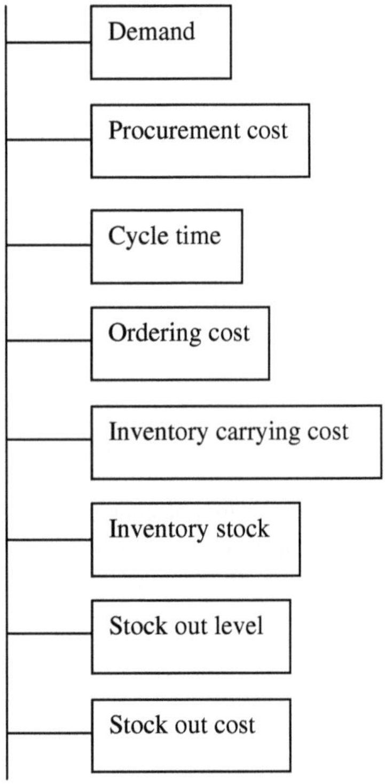

Fig. 1.3 Factors/parameters

In general, demand should be associated with certain period, i.e., the demand per unit period where the period might be:

(a) Minute
(b) Hour
(c) Day
(d) Week
(e) Month
(f) Year

Suitable period should be identified depending on the nature of business/item and sales volume of that item.

For example, if we consider only four weeks, then the approximate demands are as follows:

Week	1	2	3	4
Demand (kg)	700	700	700	700

If procurement of sugar happens in 5 kg packets, then:

$$700\,\text{kg} = 700/5 = 140\,\text{packets}$$

Therefore, the demand in packets can be expressed as follows:

Week	1	2	3	4
Demand (packets)	140	140	140	140

1.1.2 Procurement Cost

In the present context, procurement cost mainly consists of purchase cost and transportation cost as shown in Fig. 1.4.

For example, if one packet of sugar costs ₹150, then the purchase cost for week 1:

$$140 \times 150 = ₹\,21000$$

In order to carry 140 packets in suitable vehicle from wholesaler to shopkeeper, assume that the transportation cost is ₹700.

Now, the total procurement cost for week 1 is as follows:

$$21000 + 700 = ₹21700$$

In order to find out the procurement cost per unit, the total cost should be divided by the number of units procured. Therefore:

$$\text{Procurement cost per unit} = 21700 / 140$$
$$= ₹\,155$$

Breakup of the procurement cost per unit, i.e., ₹155 is as follows:

(i) Purchase cost per unit = ₹150
(ii) Transportation cost per unit = ₹5

However, in the procurement scenario from wholesaler to shopkeeper as shown in Fig. 1.5, the following situations may arise:

(a) Transportation cost is paid by the buyer
(b) Transportation is arranged by the supplier

Fig. 1.4 Components of procurement cost

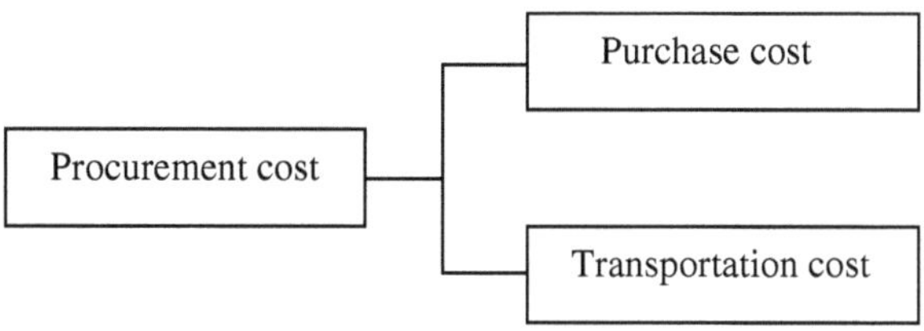

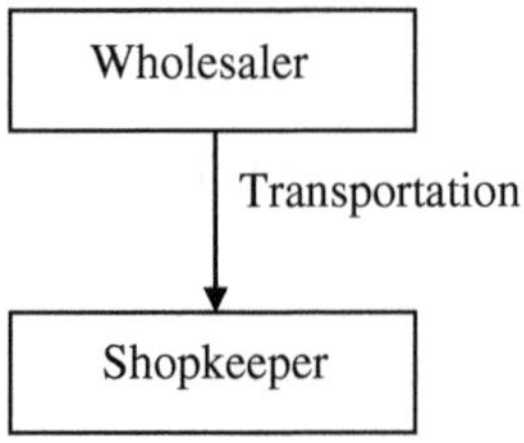

Fig. 1.5 Transportation of goods

In case where transportation cost is paid by the buyer, such costs should be included in the total procurement cost. Else the transportation and associated cost may be ignored, i.e., the price of the commodity indirectly incorporates the facility of sending the goods to the buyer location also.

1.1.3 Cycle Time

With reference to the previous example, if 700 kg or 140 packets of sugar are being procured every week, then the replenishment and consumption can be represented by Fig. 1.6.

140 packets will arrive in the beginning of week 1. After the replenishment of this quantity, the consumption or sales will happen at the demand rate of 20 packets per day or 140 packets per week approximately. Similar cycle will be repeated every week. Therefore, the cycle time is one week.

In case where the retailer procures 280 packets every two weeks, the replenishment is represented by Fig. 1.7 and the cycle time is two weeks.

1.1.4 Ordering Cost

Many shopkeepers are providing the facility of sending the goods to the end consumers as shown in Fig. 1.8. This happens after the consumer places an order

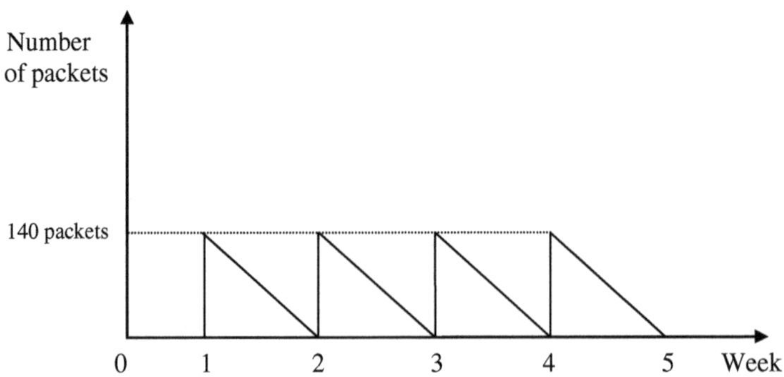

Fig. 1.6 Replenishment and consumption

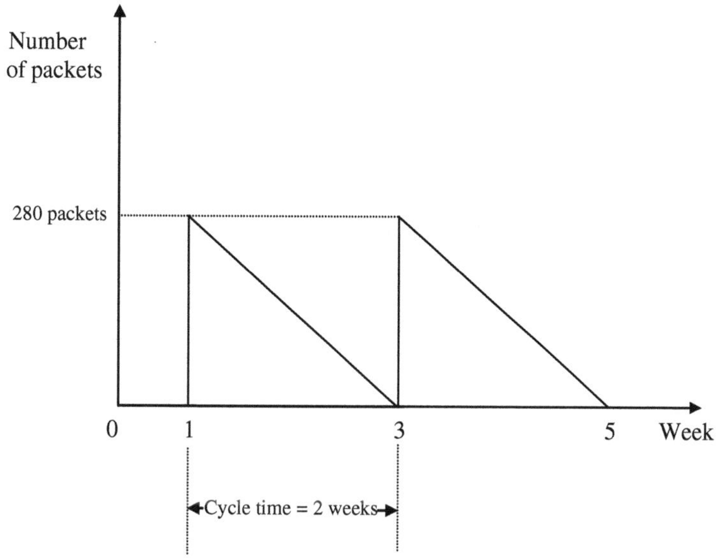

Fig. 1.7 Two weeks cycle time

Fig. 1.8 Ordering of an item

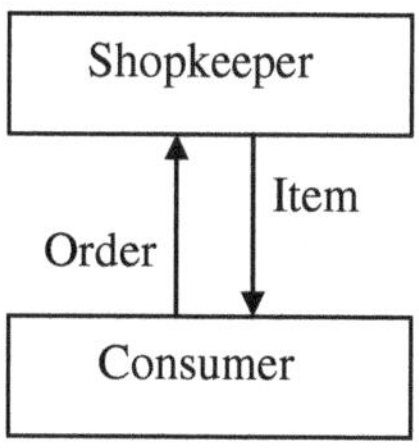

for certain item. For example, if consumer makes a telephone call from his house, then the charges associated with the call may be considered as the cost concerned with ordering. Let us say, the consumer orders for 10 kg of sugar, and ₹4 are spent in the process of ordering by the consumer. Then the ordering cost is ₹4.

Instead of 10 kg sugar, customer might order 15 kg also because of expected higher consumption. Similarly, if all the family members are expected to be out of station for almost half of a month, then the consumer might order 5 kg. The ordering quantity might vary, but the ordering cost will remain similar, i.e., ₹4 for one order. The ordering cost remains fixed per order irrespective of the ordering quantity.

In a similar manner, retailer will also place an order for wholesaler. In case of a large retail store, manager may also devote certain time on the ordering process, or some paper work might be needed. Briefly speaking, all the activities associated with the ordering process of an item should be listed, and cost concerning such activities should be estimated in order to arrive at a fixed ordering cost. Whether 140 packets of sugar are ordered or 280 packets, the ordering cost, say ₹30, will remain fixed per order.

1.1.5 Inventory Carrying Cost

As discussed before, the ordering quantity may be 140 packets or 280 packets among other options. Larger the quantity is procured; more money of the retailer will be blocked for the time being. If a smaller quantity is procured, less money is needed to purchase, and the remaining money might be in the bank as deposit. Therefore, an interest could be earned which otherwise would be lost if that money is spent in purchase of an additional quantity now.

While procuring the larger quantity, additional space is needed to store that. This additional space otherwise could be used for storing another item. With reference to Figs. 1.6 and 1.7, consider the case where 280 packets are procured instead of 140. Because of the additional quantity of 140 packets and space constraint, there is a possibility that another item (say wheat flour packet) could not be procured and stored for one week. Assume that demand rate of wheat flour packet is also similar to that of sugar. However, one packet of 5 kg wheat flour costs ₹180 (say).

For example, if 10% profit is earned by selling wheat flour packets, then the loss of profit per packet: ₹18

If sugar packet costs ₹155, then the inventory carrying cost considering only this factor (i.e., the lost opportunity for earning the profit of ₹18 per packet) is as follows:

$$18/155 = 0.116 = 11.6\%$$

For a particular period, inventory carrying cost fraction is 0.116 and carrying cost is 11.6%. In other words, carrying cost per unit item is the multiplication of inventory carrying cost fraction and the purchase/procurement cost per unit item. Therefore, the carrying cost for a particular period:

$$\begin{aligned} & 155 \times 0.116 \\ & \approx ₹\ 18 \text{ per unit item} \end{aligned}$$

Briefly speaking, the effects of higher quantity procurement should be listed and accounted in terms of money value. Finally, inventory carrying cost per unit item is estimated for certain period.

1.1.6 Inventory Stock

Inventory stock will refer to the available units at any point of time. With reference to Fig. 1.6, 140 packets are procured in the beginning of week 1. One day later, the inventory stock will be:

$$140 - 20 = 120\,\text{packets}$$

Similarly two days later, the inventory stock will be as follows:

$$140 - (2 \times 20) = 100\,\text{packets}$$

As shown in Fig. 1.9, after procurement of 140 packets, inventory stock can be found at the end of each day.

It is significant to know the stock level with respect to time so that an order can be placed timely. An idea of availability of space is helpful in planning for the procurement of some other item in case of the space constraint.

1.1.7 Stock Out Level

With reference to Fig. 1.6, 140 packets of sugar are being procured every week by the retailer. Procurement happens at the beginning of week, and sales (as per demand) happen during the week. This can be expressed as follows:

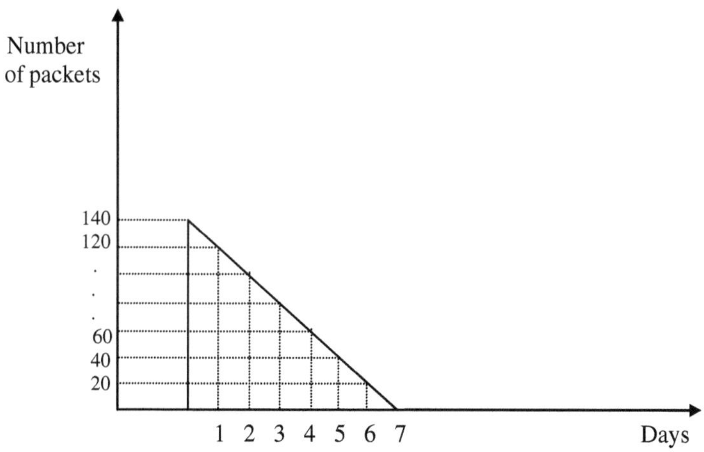

Fig. 1.9 Inventory stock with respect to time

Week	1	2	3	4
Procurement (packets)	140	140	140	140
Sales	140	140	140	140
Stock (at the end of week)	0	0	0	0

There is no stock at the end of week because demand (or sales, obviously as per the demand) is equivalent to the procurement every week. For example, assume that the demand increases to 160 in week 4 due to any reason, say festival season. Availability of packets is 140 but the demand is 160, therefore stock out level in week 4 is 20 units. In other words:

$$\begin{aligned} \text{Shortages} &= \text{Available stock} - \text{Demand} \\ &= 140 - 160 \\ &= -20 \end{aligned}$$

This is expressed as follows:

Week	1	2	3	4
Available stock	140	140	140	140
Demand	140	140	140	160
Stock (at the end of week)	0	0	0	20

There is no stock out in weeks 1, 2, and 3. But in week 4, stock out level is equivalent to 20 units.

Stock out or shortages may happen because of the following reasons:

(i) Demand exceeds supply, i.e., Demand > Supply
(ii) Demand is uniform, but supply was not on time due to some reasons.

In the previous example, demand exceeded supply, i.e., 160 instead of 140. However, as shown in Fig. 1.10, demand is uniform, i.e., 140 per week or 20 per day. But replenishment could not happen, i.e., supply is delayed by one day (say). This may be due to failure of the wholesaler to supply, such as there might be some transportation problem. Another reason for failure may be unavailability of stock at wholesaler location.

1.1.8 Stock Out Cost

If stock out situation is faced for one day, then the customer will not get such product on that day. The potential profit earned after selling that product is lost. For certain period, the stock out cost or shortage cost needs to be estimated per unit item. If sugar packet costs ₹155 and the retailer earns a profit of 10% (say), then the

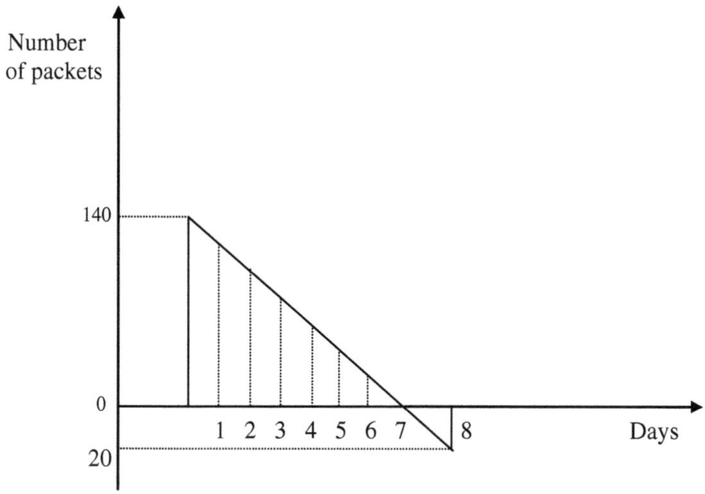

Fig. 1.10 Stock out situation

shortage cost or stock out cost for this period might be ₹15.50 per unit item. However, there may be other effects of the shortages also. The consumers (who did not get the desired item) may not visit the shop again for some time and get their requirement fulfilled from some other retailer. Therefore, the effects may prolong also.

These consumers (who are dissatisfied) may inform other known potential customers also regarding unavailability of a particular item at a particular shop. In case where the others are discouraged to visit in the near future, there is furthermore potential loss to the retailer. Briefly speaking, the effects of shortages or stock outs should be considered depending on the nature of business and the costs pertaining to that should be estimated.

1.2 Economic Lot Size

Consider the following demands:

Month/Period	1	2	3	4	5	6	7	8	9	10	11	12
Demand	50	50	50	50	50	50	50	50	50	50	50	50

Each month or period, the demand is of 50 units. Assume that the ordering cost is ₹30, and the annual inventory carrying cost is ₹10 per unit. Every month or period, the ordering quantity may not be less than 50 units in order to satisfy the demand. This is shown below:

Period	1	2	3	4	5	6	7	8	9	10	11	12
Demand	50	50	50	50	50	50	50	50	50	50	50	50
Ordering quantity	50	50	50	50	50	50	50	50	50	50	50	50
Balance	0	0	0	0	0	0	0	0	0	0	0	0

Since the twelve numbers of orders are placed, the annual or total ordering cost is as follows:

$$12 \times 30 = ₹\,360$$

Since total balance inventory is zero, the total or annual inventory carrying cost is zero.

Total annual or relevant cost = 360 + 0 = ₹360

If ordering quantity is increased to 55 (say), then this is represented as follows:

Period	1	2	3	4	5	6	7	8	9	10	11	12
Demand	50	50	50	50	50	50	50	50	50	50	50	50
Ordering quantity	55	55	55	55	55	55	55	55	55	55	50	0
Balance	5	10	15	20	25	30	35	40	45	50	50	0

Because the ordering quantity is increased, frequency of ordering or number of orders is reduced. Now eleven numbers of orders are placed and the total ordering cost is as follows:

$$11 \times 30 = ₹\,330$$

Since the annual inventory carrying cost is ₹10 per unit, this cost per period for one unit item is:

$$\frac{10}{12}$$

For period 1, the balance inventory is 5, therefore:

Carrying cost for this period

$$\text{Carrying cost for this period} = 5 \times \left(\frac{10}{12}\right)$$
$$= ₹\,4.17$$

Similarly for period 2, this is:

$$10 \times \left(\frac{10}{12}\right)$$
$$= ₹\,8.33$$

In this way, carrying cost is obtained for each period and the sum of all these will yield total inventory carrying cost.

Alternatively:

$$\text{Sum of the balance inventory} = 5 + 10 + 15 + \cdots$$
$$= 325$$

And the total carrying cost

$$\text{And the total carrying cost} = 325 \times \left(\frac{10}{12}\right)$$
$$= ₹270.83$$

While adjusting the last ordering quantity, i.e., 50 in period 11, it is also possible that it is ordered in period 12 (instead of period 11), and value of the total carrying cost will change accordingly. However, with the higher ordering quantity, carrying cost is higher.

As the ordering quantity is increased, the following effects may be observed:

(i) Total ordering cost reduces.
(ii) Total inventory carrying cost increases.

Instead if considering one type of cost alone, the focus should be on making the total cost (i.e., the sum of ordering and carrying cost) the lowest possible. With reference to Fig. 1.11, an assumption is that the consumption or demand rate is constant throughout. The ordering quantity Q arrives in the beginning, and then it is getting consumed at the rate of D units per year. In other words, the annual demand is D.

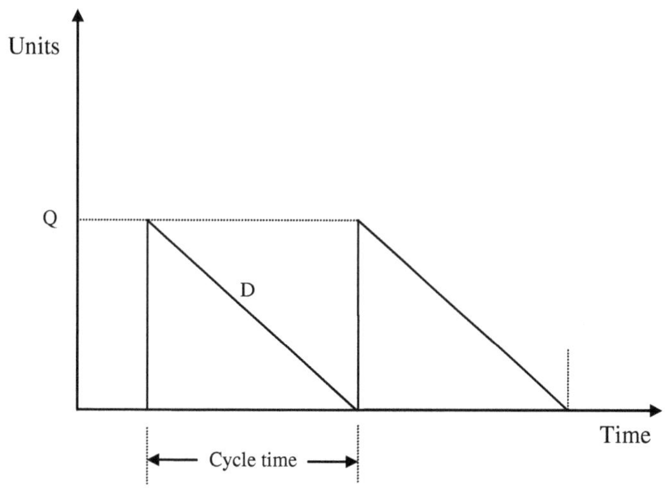

Fig. 1.11 Inventory cycle

For example, if an annual demand is 12000 units and an ordering quantity is 3000, then the number of cycles (or the number of orders) will be:

$$\frac{12000}{3000} = 4$$

Similarly if the ordering quantity is 4000, then the number of orders will be:

$$\frac{12000}{4000} = 3$$

In order to generalize, the number of orders:

$$\frac{D}{Q}$$

If the fixed ordering cost per unit order is C, then the annual ordering cost (AOC) is as follows:

$$AOC = \left[\frac{D}{Q}\right].C \tag{1.1}$$

In the beginning of the cycle time, available inventory is Q units, and it is zero at the end of the cycle time. Such inventory cycle continues throughout the year, therefore the average inventory is:

$$\frac{Q}{2}$$

If the annual inventory carrying cost is I per unit item, then the annual inventory carrying cost (AIC) is as follows:

$$AIC = \left[\frac{Q}{2}\right].I \tag{1.2}$$

Now the total annual cost (E) can be obtained by adding Eqs. (1.1) and (1.2), i.e.,

$$E = AOC + AIC$$

Or

$$E = \left[\frac{D}{Q}\right].C + \left[\frac{Q}{2}\right].I \tag{1.3}$$

Since the objective is to minimize the total cost, differentiate the Eq. (1.3) with respect to Q and make it equivalent to zero:

$$\frac{-DC}{Q^2} + \frac{I}{2} = 0$$

Or

$$\frac{DC}{Q^2} = \frac{I}{2}$$

Or

$$Q^2 = \frac{2DC}{I}$$

Therefore, the economic lot size:

$$Q^* = \sqrt{\frac{2DC}{I}} \tag{1.4}$$

Substituting Q^* in Eq. (1.3), optimal total cost:

$$E^* = DC.\sqrt{\frac{I}{2DC}} + \frac{I}{2}.\sqrt{\frac{2DC}{I}}$$

Or

$$E^* = \sqrt{\frac{DCI}{2}} + \sqrt{\frac{DCI}{2}}$$

[It can be observed that both the components of the E^* are equal, i.e., $\sqrt{\frac{DCI}{2}}$. Therefore, the economic lot size for such situation refers to the ordering quantity Q^* corresponding to which, the annual ordering cost is equivalent to the annual inventory carrying cost.]

$$E^* = 2\sqrt{\frac{DCI}{2}}$$

Or

$$E^* = \sqrt{2DCI} \tag{1.5}$$

In case where stock out situation is incorporated, it is represented by Fig. 1.12.

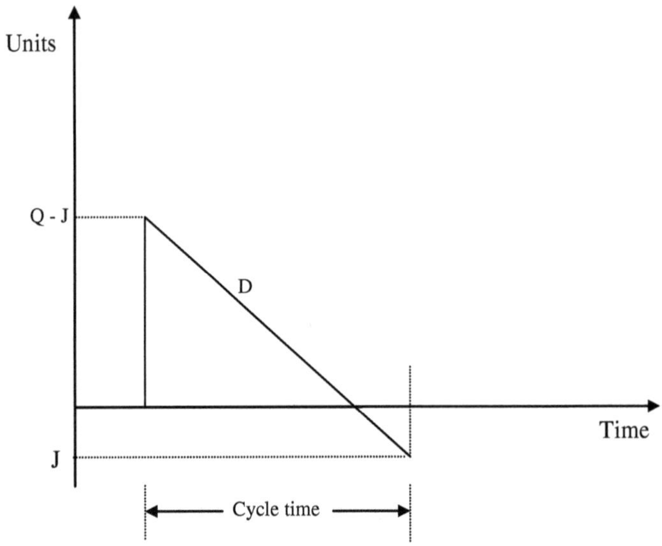

Fig. 1.12 Inventory cycle with stock out

The cycle time includes stock out time and this cycle continues throughout. The stock out is only for a part of cycle time. For example, if stock out is for two days and constant demand per day is 50, then the maximum stock out or shortages will be:

$$50 \times 2 = 100 \text{ units}$$

Assume the following additional notation:

J = Maximum stock out or shortages
K = Annual stock out cost per unit

The stock out situation is for the time:

$$\frac{J}{D}$$

The average stock out is as follows:

$$\frac{J}{2}$$

Since there are $\frac{D}{Q}$ cycles in a year, the annual stock out cost (ASC) is given as:

$$\frac{J}{2} \cdot \frac{J}{D} . K . \frac{D}{Q}$$

Or

$$ASC = \frac{J^2K}{2Q} \tag{1.6}$$

As all the stock out units are assumed to be back-ordered, the procurement ordering quantity Q suitably incorporates J. Therefore, the positive inventory exists for the time:

$$\frac{(Q-J)}{D}$$

Average positive inventory is $\frac{(Q-J)}{2}$ and there are $\frac{D}{Q}$ cycles. The annual inventory carrying cost is expressed as:

$$AIC = \frac{(Q-J)}{2} \cdot \frac{(Q-J)}{D} . I . \frac{D}{Q}$$

Or

$$AIC = \frac{(Q-J)^2 I}{2Q} \tag{1.7}$$

Annual ordering cost is similar to the Eq. (1.1). Adding Eqs. (1.6), (1.7), and (1.1), total annual cost:

$$E = \frac{J^2K}{2Q} + \frac{(Q-J)^2 I}{2Q} + \frac{DC}{Q}$$

Or

$$E = \frac{J^2K}{2Q} + \frac{(Q^2 - 2QJ + J^2)I}{2Q} + \frac{DC}{Q}$$

Or

$$E = \frac{J^2K}{2Q} + \frac{QI}{2} - JI + \frac{J^2I}{2Q} + \frac{DC}{Q}$$

Or

$$E = \frac{QI}{2} + \frac{DC}{Q} + \frac{J^2(K+I)}{2Q} - JI \tag{1.8}$$

In addition to the economic lot size, an optimum value of maximum stock out units is also to be determined. Differentiating Eq. (1.8) partially with respect to J and equating to zero:

$$\frac{J(K+I)}{Q} - I = 0$$

Or

$$J = \frac{IQ}{(K+I)} \tag{1.9}$$

Substituting value of J in Eq. (1.8):

$$E = \frac{QI}{2} + \frac{DC}{Q} + \frac{I^2Q^2}{(K+I)^2} \cdot \frac{(K+I)}{2Q} - \frac{I^2Q}{(K+I)}$$

Or

$$E = \frac{QI}{2} + \frac{DC}{Q} + \frac{I^2Q}{2(K+I)} - \frac{I^2Q}{(K+I)}$$

Or

$$E = \frac{QI}{2} + \frac{DC}{Q} - \frac{I^2Q}{2(K+I)}$$

Or

$$E = \frac{QIK}{2(K+I)} + \frac{DC}{Q} \tag{1.10}$$

Differentiating with respect to Q and equating to zero:

$$\frac{IK}{2(K+I)} - \frac{DC}{Q^2} = 0$$

Or

$$\frac{IK}{2(K+I)} = \frac{DC}{Q^2}$$

Or

$$Q^2 = \frac{2DC(K+I)}{IK}$$

Or

$$Q^* = \sqrt{\frac{2DC(K+I)}{IK}} \tag{1.11}$$

Substituting this optimal lot size Q^* (for the stock out scenario) in Eq. (1.9), an optimal value of maximum stock out units:

$$J^* = \frac{I}{(K+I)}\sqrt{\frac{2DC(K+I)}{IK}}$$

Or

$$J^* = \sqrt{\frac{2DCI}{K(K+I)}} \tag{1.12}$$

Substituting optimal values of Q and J from Eqs. (1.11) and (1.12), respectively, into Eq. (1.8), the optimal total cost is derived as:

$$E = \frac{I}{2}\sqrt{\frac{2DC(K+I)}{IK}} + DC.\sqrt{\frac{IK}{2DC(K+I)}} - J\left[I - \frac{J(K+I)}{2Q}\right]$$

Or

$$E = \sqrt{\frac{DCI(K+I)}{2K}} + \sqrt{\frac{DCIK}{2(K+I)}} - J\left[I - \frac{J(K+I)}{2Q}\right]$$

Or

$$E = \sqrt{\frac{DCI}{2K(K+I)}}[(K+I)+K] - J\left[I - \frac{J(K+I)}{2Q}\right]$$

Or

$$E = \sqrt{\frac{DCI}{2K(K+I)}}[2K+I] - \sqrt{\frac{2DCI}{K(K+I)}}\left[I - \frac{(K+I)}{2}\sqrt{\frac{2DCI}{K(K+I)}}\sqrt{\frac{IK}{2DC(K+I)}}\right]$$

Or

$$E = \sqrt{\frac{DCI}{2K(K+I)}}[2K+I] - \sqrt{\frac{2DCI}{K(K+I)}}\left[I - \frac{I}{2}\right]$$

Or

$$E = \sqrt{\frac{DCI}{2K(K+I)}}[2K+I] - \sqrt{\frac{DCI}{2K(K+I)}}[I]$$

Or

$$E = \sqrt{\frac{DCI}{2K(K+I)}}[2K+I-I]$$

Or

$$E = 2K\sqrt{\frac{DCI}{2K(K+I)}}$$

Or

$$E^* = \sqrt{\frac{2DCIK}{(K+I)}} \tag{1.13}$$

In this chapter, various parameters concerning the inventory management are discussed. In the context of economic lot size, the equations related to the following optimum values have been provided:

(i) Optimum lot size
(ii) Optimum total relevant cost

Additionally while incorporating the stock out situation, the equations concerning an optimum value of stock out quantity are provided along with lot size and total cost. These equations will be used for furthermore analysis related to various parameters in addition to other details.

1.3 Interaction of Inventory Parameters

As it is derived before, the following inventory parameters are considered for the case without stock outs:

(a) Demand
(b) Ordering cost
(c) Inventory holding cost

Interaction among these parameters is also of interest concerning the relevant decisions.

1.3.1 Basic Case Without Stock Outs

When demand increases, a procurement lot size and the related cost increase. Because of this demand increase, profit earned by the company is higher. However, a change in the optimum procurement lot size might not be preferred by the organization because of the inconvenience (or it is not so economic) in transportation/material handling. Therefore, a conscious effort can be made to keep the lot size similar. If it is possible, an ordering cost may be reduced for achieving such an objective. It is of interest to determine downward variation in ordering cost corresponding to an upward variation in demand.

With the demand increase, the total optimal cost also increases. In case where the management wants to maintain similar total cost as before, then either the ordering or inventory holding cost might be reduced. However, convenience in implementing such cost reduction should be visualized. For a company, if there is no chance of furthermore ordering cost decrease, then the management might explore the holding cost decrease. Where it is not feasible to go for holding cost reduction any more, the ordering cost decrease might be explored. In case where both the parameters may be reduced practically, the potential amount of reduction might be at a lower scale in comparison with single-parameter decrease.

In general also, conscious efforts are made for an increase in demand and the decrease in cost parameters such as ordering and holding cost. However, any parameter may vary in a dynamic business environment and that triggers the management response in terms of another suitable parameter variation with an objective such as:

(a) Similar lot size
(b) Similar total cost

Table 1.1 shows the possible management response when the aim is to have similar lot size, i.e., as it was before the change is triggered by certain parameter.

With the aim of similar total cost, possible management response is shown in Table 1.2.

Table 1.1 Management response with the aim of similar lot size

Change triggered by	Response
Increase in demand	Reduction in ordering cost
Reduction in demand	Reduction in inventory holding cost
Ordering cost decrease	Potential demand increase/holding cost reduction
Holding cost reduction	Ordering cost reduction
Increase in holding cost	Potential demand increase

Table 1.2 Management response with the aim of similar total cost

Change triggered by	Response
Holding cost increase	Reduction in ordering cost
Reduction in holding cost	Potential demand increase
Increase in ordering cost	Holding cost reduction
Reduction in ordering cost	Potential demand increase
Demand increase	Reduction in ordering/carrying cost

1.3.2 With Stock Out

The following inventory parameters are considered for the case with an inclusion of the shortages:

(a) Ordering cost
(b) Inventory carrying cost
(c) Demand
(d) Stock out cost

Interaction among inventory parameters needs to be analyzed as these parameters may vary. As mentioned before, management response is in accordance with the following objective:

(i) Similar lot size
(ii) Similar total cost

Table 1.3 represents the response of management for similar lot size. In certain cases, similar lot size is preferred by relevant stakeholders. Therefore, under stock out situation also, a response can be planned in view of change triggered by certain parameter in the context of lot size. When a suitable response consists of more than one option, either option can be considered depending on the convenience in implementing it. Else a combination of different options may be evaluated qualitatively and quantitatively and might be applied depending on an ease in

Table 1.3 Response in case of stock out for similar lot size

Change triggered by	Response
Ordering cost reduction	Demand increase/carrying cost reduction/stock out cost reduction
Decrease in carrying cost	Ordering cost reduction
Increase in carrying cost	Demand increase/stock out cost reduction
Demand increase	Ordering cost reduction
Demand reduction	Carrying cost reduction/stock out cost reduction
Stock out cost increase	Demand increase/carrying cost reduction
Stock out cost decrease	Ordering cost reduction

Table 1.4 Response in case of stock out for similar total cost

Change triggered by	Response
Decrease in ordering cost	Demand increase
Increase in ordering cost	Carrying cost reduction/stock out cost reduction
Carrying cost reduction	Demand increase
Carrying cost increase	Ordering cost reduction/stock out cost reduction
Increase in demand	Ordering cost decrease/carrying cost decrease/stock out cost decrease
Stock out cost reduction	Demand increase
Stock out cost increase	Ordering cost reduction/carrying cost reduction

implementation in a particular business/operational setting. Additional factors may also be considered pertaining to variation of different parameters before taking a final decision.

Potential response of the management is shown in Table 1.4 for an objective of attaining the similar total cost.

Chapter 2
Demand

Demand needs to be estimated for the future periods. However, demand can fluctuate also. Variation in the demand might be as follows:

(i) Increase in demand
(ii) Decrease in demand

An increase in demand may be owing to the reasons as follows:

(a) The product may be liked by the customers, and therefore the demand has increased. Either the number of customers might increase or similar number of consumers started purchasing the item in comparatively larger numbers. It may also be possible that number of customers and the purchased items per customer both have increased.
(b) Because of the competition with other producer companies, the price per item is reduced. The consumers have started purchasing larger quantities due to comparatively lower price.
(c) As the product is advertised a lot, the sales have increased.

Demand rate increase can be depicted by Fig. 2.1, where annual demand or demand rate per year, D is increased to D_1. Because of higher demand rate, the similar lot will be consumed earlier.

Similarly, a decrease in demand might be because of the following reasons among others:

(a) The product is not liked by the consumers. Either the number of customers has decreased or similar number of consumers started purchasing in smaller quantities. There may also be a possibility that number of customers and the desired items per customer both have decreased.
(b) For the reasons, such as an increased cost of transportation and/or manufacturing, price of the item has increased. Because of an increased price, customers might purchase in lower quantities.

S. Sharma, *Inventory Parameters*, DOI 10.1007/978-981-10-4545-5_2

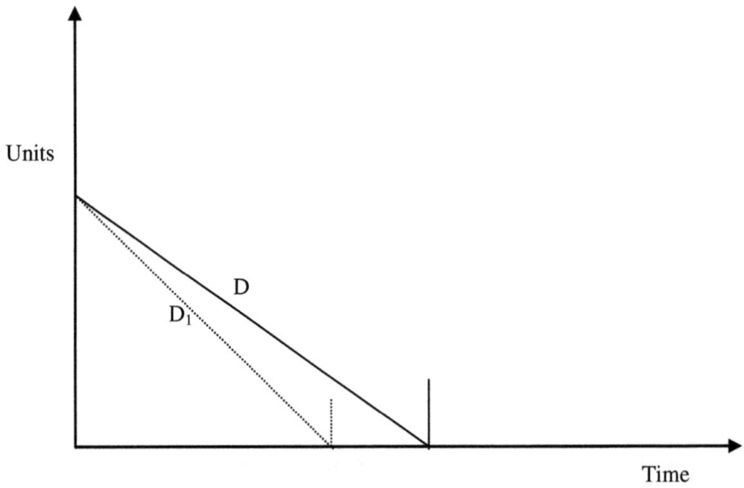

Fig. 2.1 Demand rate increase

(c) The sales have decreased because of other substitutes/alternatives available in the market.

Demand rate decrease can be depicted by Fig. 2.2, where D is decreased to D_1. Because of a decreased demand rate, the similar lot will be completely consumed taking comparatively longer time.

Demand and its variation are among significant factors to be considered for analysis concerning inventory management.

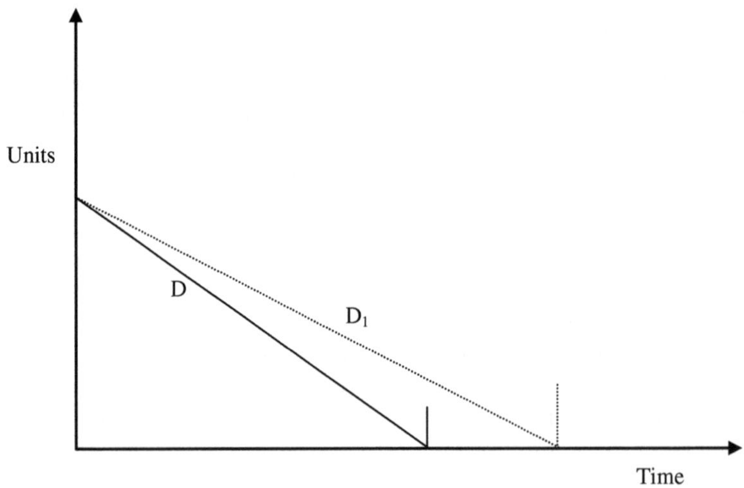

Fig. 2.2 Demand rate decrease

2.1 Increase in Demand

With an increase in demand, the profit earned by the retailer is advantageous. However, it is necessary to know the effects on the purchasing pattern and also on the space requirement to store the items.

Example 2.1
Assume the following parameters:

Annual demand, D = 600 units
Ordering cost, C = ₹30
Annual inventory carrying cost per unit, I = ₹10

Using the Eq. (1.4),

$$Q^* = \sqrt{\frac{2DC}{I}} = \sqrt{\frac{2 \times 600 \times 30}{10}} = 60 \text{ units}$$

And, the total related annual cost from Eq. (1.5),

$$E^* = \sqrt{2DCI} = \sqrt{2 \times 600 \times 30 \times 10} = ₹600$$

Now, if demand rate is increased by 15% (say), then the new annual demand is 690 units. The calculation for a revised purchasing lot size and cost is as follows:

$$Q^* = \sqrt{\frac{2 \times 690 \times 30}{10}} = 64.34 \approx 64 \text{ units}$$

And

$$E^* = \sqrt{2 \times 690 \times 30 \times 10} = ₹643.43 \approx ₹643$$

Instead of purchasing 60 units as lot size, 64 units are now procured in one lot.

$$\text{Relative change in } Q = \frac{64.34 - 60}{60} = \frac{4.34}{60} = 0.072$$

$$\% \text{ change in } Q = 0.072 \times 100 = 7.2\%$$

$$\text{Similarly \% change in } E = \frac{643.43 - 600}{600} = \frac{43.43}{600} = 0.072$$
$$= 7.2\%$$

Example 2.2
Consider the base data of Example 2.1. Analyze if an increase in D is as follows:

Increase in D	5%	10%	15%	20%	25%	30%
D	630	660	690	720	750	780

Following the procedure explained in Example 2.1, demand D is increased by the given percentage and approximate values are obtained such as:

(i) Q
(ii) % increase in Q
(iii) E
(iv) % increase in E

Table 2.1 represents the variation in parameters with reference to an increase in demand.

With the increase in demand, optimum lot size and related cost increase. Such information helps in the following activities:

(i) Procurement planning
(ii) Transportation planning
(iii) Arrangement of space required
(iv) Related costs/expenditure in purchase of larger quantities

Let:

P = % variation in parameter

For the present discussion, P refers to the % increase in demand, therefore the increased demand:

Table 2.1 Effects on parameters with respect to demand increase

Increase in D	5%	10%	15%	20%	25%	30%
I	630	660	690	720	750	780
Q	61	63	64	66	67	68
% Increase in Q	2.5%	4.9%	7.2%	9.5%	11.8%	14.0%
E	615	629	643	657	671	684
% Increase in E	2.5%	4.9%	7.2%	9.5%	11.8%	14.0%

$$D_1 = \left(1 + \frac{P}{100}\right) D$$

$$\begin{aligned}
\text{Increase in the lot size} &= \sqrt{\frac{2D_1C}{I}} - \sqrt{\frac{2DC}{I}} \\
&= \sqrt{\frac{2\left(1 + \frac{P}{100}\right)DC}{I}} - \sqrt{\frac{2DC}{I}} \\
&= \sqrt{\frac{2DC}{I}}\left[\sqrt{\left(1 + \frac{P}{100}\right)} - 1\right]
\end{aligned}$$

And:

$$\%\text{ increase in } Q = \sqrt{\left(1 + \frac{P}{100}\right)} - 1$$

Now:

$$\begin{aligned}
\text{Additional cost incurred} &= \sqrt{2D_1CI} - \sqrt{2DCI} \\
&= \sqrt{2CI}\left[\sqrt{D_1} - \sqrt{D}\right] \\
&= \sqrt{2CI}\left[\sqrt{\left(1 + \frac{P}{100}\right)D} - \sqrt{D}\right] \\
&= \sqrt{2DCI}\left[\sqrt{\left(1 + \frac{P}{100}\right)} - 1\right]
\end{aligned}$$

And:

$$\%\text{ increase in } E = \sqrt{\left(1 + \frac{P}{100}\right)} - 1$$

These generalized results help the decision makers/business owners to know well in advance the estimation related to the following:

(a) What will be the increase in optimum lot size?
(b) How much additional space would be needed to place the additional units of an item?
(c) How much additional expenditure would be required because of the cost increase?

The derived results are also summarized in Table 2.2.

Table 2.2 Results with reference to % increase in demand

Increase in the lot size	$\sqrt{\frac{2DC}{I}}\left[\sqrt{\left(1+\frac{P}{100}\right)}-1\right]$
% Increase in the lot size	$\sqrt{\left(1+\frac{P}{100}\right)}-1$
Additional related cost	$\sqrt{2DCI}\left[\sqrt{\left(1+\frac{P}{100}\right)}-1\right]$
% Increase in cost	$\sqrt{\left(1+\frac{P}{100}\right)}-1$

2.2 Decrease in Demand

Although a decrease in demand is not preferred in general, yet there is every possibility to face it. With a decrease in demand, there is profit loss. However, that can be overcome by some other item. It is of practical relevance to know the effects of a decrease in demand on various parameters.

Example 2.3
Consider the base data of Example 2.1. Analyze whether a decrease in D is as follows:

% Decrease in D	5%	10%	15%	20%	25%	30%
D	570	540	510	480	450	420

Following the procedure explained in Example 2.1, demand D is decreased by the given percentage and approximate values are obtained such as:

(i) Q
(ii) % decrease in Q
(iii) E
(iv) % decrease in E

Table 2.3 represents the approximate variation in parameters with reference to a decrease in demand.

With the decrease in demand, optimum lot size and related cost decrease. However, the % decrease is more pronounced in comparison with the previous situation, i.e., the % variation of output parameters corresponding to the demand increase. Such information helps in the following activities:

Table 2.3 Effects on parameters with respect to demand decrease

% Decrease in D	5%	10%	15%	20%	25%	30%
D	570	540	510	480	450	420
Q	58	57	55	54	52	50
% Decrease in Q	2.5%	5.1%	7.8%	10.6%	13.4%	16.3%
E	585	569	553	537	520	502
% Decrease in E	2.5%	5.1%	7.8%	10.6%	13.4%	16.3%

(i) Purchase planning
(ii) Effects on material handling
(iii) How much space would be vacant?
(iv) How much fund is expected to be available for alternate investment?

Let:

P = % decrease in demand

The decreased demand:

$$D_1 = \left(1 - \frac{P}{100}\right)D$$

$$\begin{aligned} \text{Decrease in the lot size } &= \sqrt{\frac{2DC}{I}} - \sqrt{\frac{2D_1C}{I}} \\ &= \sqrt{\frac{2DC}{I}} - \sqrt{\frac{2(1 - \frac{P}{100})DC}{I}} \\ &= \sqrt{\frac{2DC}{I}}\left[1 - \sqrt{\left(1 - \frac{P}{100}\right)}\right] \end{aligned}$$

And:

$$\% \text{ decrease in } Q = 1 - \sqrt{\left(1 - \frac{P}{100}\right)}$$

Now:

$$\begin{aligned} \text{Cost reduction} &= \sqrt{2DCI} - \sqrt{2D_1CI} \\ &= \sqrt{2CI}\left[\sqrt{D} - \sqrt{D_1}\right] \\ &= \sqrt{2CI}\left[\sqrt{D} - \sqrt{\left(1 - \frac{P}{100}\right)D}\right] \\ &= \sqrt{2DCI}\left[1 - \sqrt{\left(1 - \frac{P}{100}\right)}\right] \end{aligned}$$

And:

$$\% \text{ decrease in } E = 1 - \sqrt{\left(1 - \frac{P}{100}\right)}$$

Table 2.4 Results with reference to % decrease in demand

Decrease in the lot size	$\sqrt{\frac{2DC}{I}}\left[1-\sqrt{\left(1-\frac{P}{100}\right)}\right]$
% Decrease in the lot size	$1-\sqrt{\left(1-\frac{P}{100}\right)}$
Cost reduction	$\sqrt{2DCI}\left[1-\sqrt{\left(1-\frac{P}{100}\right)}\right]$
% Decrease in cost	$1-\sqrt{\left(1-\frac{P}{100}\right)}$

These generalized results help the decision makers/business owners to know well in advance the estimation related to the following:

(a) What will be the decrease in optimum lot size?
(b) How much additional space would be available to place some other inventory item?
(c) In order to compensate for the loss of profit, how much additional money is available for purchase, store, and sale of some other suitable item?

The derived results are also summarized in Table 2.4.

2.3 Stock Out Situation

The demand exists in the stock out situation, but there is no inventory stock to satisfy those demands. This is shown in Fig. 2.3.

Example 2.4
Assume the following parameters:

Annual demand, D = 600 units
Ordering cost, C = ₹30
Annual inventory carrying cost per unit, I = ₹10
Annual stock out cost per unit, K = ₹200

Using the Eq. (1.11),

$$Q^* = \sqrt{\frac{2DC(K+I)}{IK}}$$
$$= \sqrt{\frac{2\times 600\times 30(200+10)}{10\times 200}}$$
$$= 61.4817 \text{ units}$$

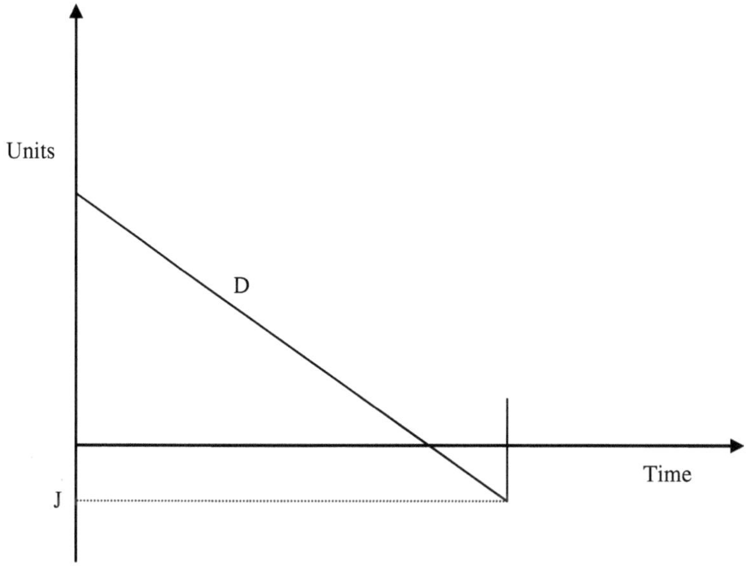

Fig. 2.3 Stock out situation

From the Eq. (1.12), optimum stock out units:

$$J^* = \sqrt{\frac{2DCI}{K(K+I)}}$$
$$= \sqrt{\frac{2 \times 600 \times 30 \times 10}{200(200+10)}}$$
$$= 2.93 \text{ units}$$

And, the total related annual cost from Eq. (1.13),

$$E^* = \sqrt{\frac{2DCIK}{(K+I)}}$$
$$= \sqrt{\frac{2 \times 600 \times 30 \times 10 \times 200}{(200+10)}}$$
$$= ₹585.54$$

If the stock out situation is properly managed, then there is an overall reduced total cost by optimizing the procurement ordering quantity and maximum stock out units. In order to handle the stock out scenario, the ordering quantity has relatively increased. Thus, the introduced shortages or stock outs (in comparison with no stock outs) result into the following:

(i) An increased order size
(ii) A decreased total related cost

Such quantification helps in knowing well in advance the requirement of additional space and also the availability of additional money for better potential use.

2.3.1 Demand Increase

Assume that the demand D is increased to D_1. This is shown in Fig. 2.4. Since a variation in demand affects the order size, stock out units, and total cost, it is of interest to analyze these effects. As it is explained before, the demand is increased by certain percentage and analysis is conducted after computation.

Example 2.5
Consider the base data of Example 2.4 as given below:

D	C	I	K	Q	E	J
600	30	10	200	61.4817	585.54	2.93

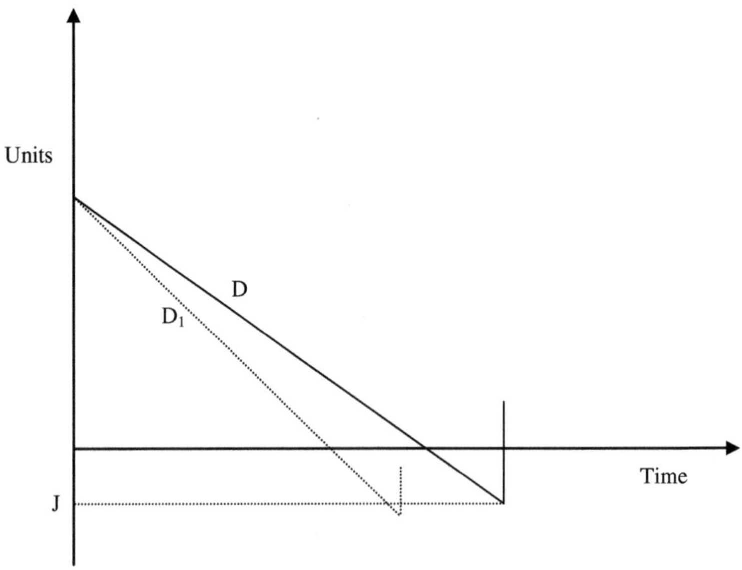

Fig. 2.4 Demand increase with stock out situation

Table 2.5 Effects on parameters with stock out corresponding to demand increase

% Increase in D	5%	10%	15%	20%	25%	30%
D	630	660	690	720	750	780
Q	63	64	66	67	69	70
% Increase in Q	2.5%	4.9%	7.2%	9.5%	11.8%	14.0%
E	600	614	628	641	655	668
% Increase in E	2.5%	4.9%	7.2%	9.5%	11.8%	14.0%
J	3.00	3.07	3.14	3.21	3.27	3.34
% Increase in J	2.4%	4.8%	7.2%	9.5%	11.7%	13.9%

Analyze if an increase in D is as follows:

% Increase in D	5%	10%	15%	20%	25%	30%
D	630	660	690	720	750	780

Following the procedure explained in Example 2.4, demand D is increased by the given percentage and approximate values are obtained such as:

(i) Q
(ii) % increase in Q
(iii) E
(iv) % increase in E
(v) J
(vi) % increase in J

Table 2.5 represents the approximate variation in parameters with reference to an increase in demand. Optimum lot size and cost increase. The analysis helps in knowing the effects on optimum value of maximum stock out units that tend to increase. Usefulness also lies in setting suitable value of maximum shortages that are allowed and planning for other operational activities (on the basis of the output parameters thus arrived), after imposing various practical realities specific to a particular business and consumer/customer group. A successful implementation depends on the estimation of stock out cost for a particular item/business/industry.

As an increased demand $D_1 = \left(1 + \frac{P}{100}\right)D$, and lot size $Q^* = \sqrt{\frac{2DC(K+I)}{IK}}$, increase in lot size:

$$= \sqrt{\frac{2\left(1 + \frac{P}{100}\right)DC(K+I)}{IK}} - \sqrt{\frac{2DC(K+I)}{IK}}$$

$$= \sqrt{\frac{2DC(K+I)}{IK}}\left[\sqrt{\left(1 + \frac{P}{100}\right)} - 1\right]$$

$$\text{Increase in the optimum stock out units} = \sqrt{\frac{2D_1CI}{K(K+I)}} - \sqrt{\frac{2DCI}{K(K+I)}}$$

$$= \sqrt{\frac{2\left(1+\frac{P}{100}\right)DCI}{K(K+I)}} - \sqrt{\frac{2DCI}{K(K+I)}}$$

$$= \sqrt{\frac{2DCI}{K(K+I)}}\left[\sqrt{\left(1+\frac{P}{100}\right)} - 1\right]$$

And, increase in the total related cost:

$$= \sqrt{\frac{2D_1CIK}{(K+I)}} - \sqrt{\frac{2DCIK}{(K+I)}}$$

$$= \sqrt{\frac{2DCIK}{(K+I)}}\left[\sqrt{\left(1+\frac{P}{100}\right)} - 1\right]$$

The derived results are also summarized in Table 2.6.

2.3.2 *Demand Decrease*

Assume that the demand D is decreased to D_1. This is shown in Fig. 2.5 as the decreased demand takes longer time to consume.

Example 2.6

Consider the base data of Example 2.4 as given below:

D	C	I	K	Q	E	J
600	30	10	200	61.48	585.54	2.93

Table 2.6 Results with reference to % increase in demand along with stock out

Increase in the lot size	$\sqrt{\frac{2DC(K+I)}{IK}}\left[\sqrt{\left(1+\frac{P}{100}\right)} - 1\right]$
% Increase in the lot size	$\sqrt{\left(1+\frac{P}{100}\right)} - 1$
Additional related cost	$\sqrt{\frac{2DCIK}{(K+I)}}\left[\sqrt{\left(1+\frac{P}{100}\right)} - 1\right]$
% Increase in cost	$\sqrt{\left(1+\frac{P}{100}\right)} - 1$
Increase in the stock out units	$\sqrt{\frac{2DCI}{K(K+I)}}\left[\sqrt{\left(1+\frac{P}{100}\right)} - 1\right]$
% Increase in the stock out units	$\sqrt{\left(1+\frac{P}{100}\right)} - 1$

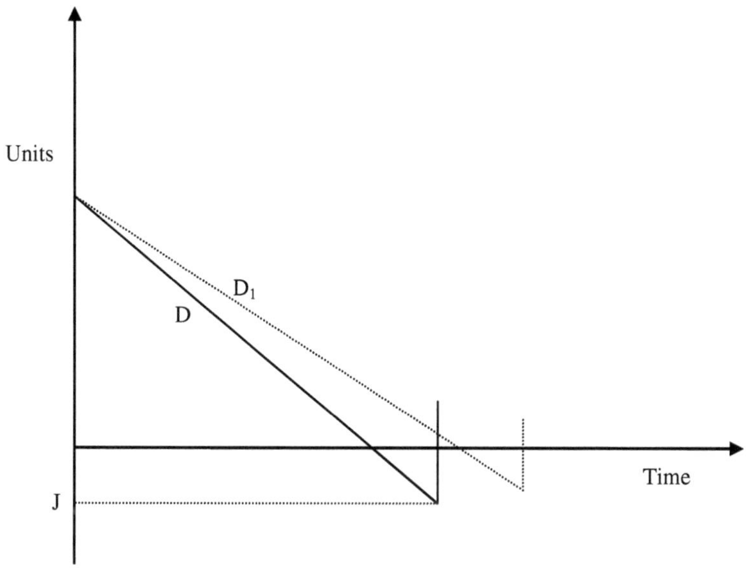

Fig. 2.5 Demand decrease with stock out situation

Analyze if a decrease in *D* is as follows:

% Decrease in *D*	5%	10%	15%	20%	25%	30%
D	570	540	510	480	450	420

Demand *D* is now decreased by the given percentage and approximate values are obtained such as:

(i) *Q*
(ii) % decrease in *Q*
(iii) *E*
(iv) % decrease in *E*
(v) *J*
(vi) % decrease in *J*

Table 2.7 represents the variation in parameters with reference to a decrease in demand, Optimum lot size, and cost decrease. The analysis helps in knowing the effects on optimum value of maximum stock out units that tend to decrease. Various resources including space and money will be available for alternative use.

A reduction in demand and its effects are illustrated with the use of example. However, it is necessary to generalize the results for wide applications concerning any business/industry/product type. Such a generalization of the results will help in developing furthermore insights.

Table 2.7 Effects on parameters with stock out corresponding to demand decrease

% Decrease in D	5%	10%	15%	20%	25%	30%
D	570	540	510	480	450	420
Q	59.92	58.33	56.68	54.99	53.24	51.44
% Decrease in Q	2.53%	5.13%	7.80%	10.56%	13.40%	16.33%
J	2.85	2.78	2.70	2.62	2.54	2.45
% Decrease in J	2.53%	5.13%	7.80%	10.56%	13.40%	16.33%
E	570.71	555.49	539.84	523.72	507.09	489.90
% Decrease in E	2.53	5.13	7.80	10.56	13.40	16.33

Let:

P = % decrease in demand

The decreased demand:

$$D_1 = \left(1 - \frac{P}{100}\right)D$$

Decrease in the lot size:

$$= \sqrt{\frac{2DC(K+I)}{IK}} - \sqrt{\frac{2\left(1 - \frac{P}{100}\right)DC(K+I)}{IK}}$$

$$= \sqrt{\frac{2DC(K+I)}{IK}}\left[1 - \sqrt{\left(1 - \frac{P}{100}\right)}\right]$$

$$\text{Decrease in the optimum stock out units} = \sqrt{\frac{2DCI}{K(K+I)}} - \sqrt{\frac{2D_1CI}{K(K+I)}}$$

$$= \sqrt{\frac{2DCI}{K(K+I)}} - \sqrt{\frac{2\left(1 - \frac{P}{100}\right)DCI}{K(K+I)}}$$

$$= \sqrt{\frac{2DCI}{K(K+I)}}\left[1 - \sqrt{\left(1 - \frac{P}{100}\right)}\right]$$

And; decrease in the total related cost:

$$= \sqrt{\frac{2DCIK}{(K+I)}} - \sqrt{\frac{2D_1CIK}{(K+I)}}$$

$$= \sqrt{\frac{2DCIK}{(K+I)}}\left[1 - \sqrt{\left(1 - \frac{P}{100}\right)}\right]$$

Decrease in the demand results into:

(a) Reduction in the procurement lot size
(b) Decrease in the optimum stock out units
(c) Reduction in the total related cost

As discussed before, such information is useful for the estimation of the following:

(i) Reduction in the applicable transportation and material handling cost
(ii) Availability of space for storing and selling other alternative item
(iii) Availability of funds for a particular business situation

The derived results are also summarized in Table 2.8.

% decrease in the output parameters (concerning demand reduction)

$$= 1 - \sqrt{\left(1 - \frac{P}{100}\right)}$$

% increase in the output parameters (concerning demand increase)

$$= \sqrt{\left(1 + \frac{P}{100}\right)} - 1$$

The percentage decrease in the output parameters is relatively more in comparison with the percentage increase. This is demonstrated in Table 2.9. For all practical purposes, the demand cannot decrease by more than 100%, i.e., $P \leq 100$.

Analytically also, it can be shown for $P \leq 100$:

$$1 - \sqrt{\left(1 - \frac{P}{100}\right)} > \sqrt{\left(1 + \frac{P}{100}\right)} - 1$$

Table 2.8 Results with reference to % decrease in demand along with stock out

Decrease in the lot size	$\sqrt{\frac{2DC(K+I)}{IK}}\left[1 - \sqrt{\left(1 - \frac{P}{100}\right)}\right]$
% Decrease in the lot size	$1 - \sqrt{\left(1 - \frac{P}{100}\right)}$
Cost reduction	$\sqrt{\frac{2DCIK}{(K+I)}}\left[1 - \sqrt{\left(1 - \frac{P}{100}\right)}\right]$
% Decrease in cost	$1 - \sqrt{\left(1 - \frac{P}{100}\right)}$
Decrease in the stock out units	$\sqrt{\frac{2DCI}{K(K+I)}}\left[1 - \sqrt{\left(1 - \frac{P}{100}\right)}\right]$
% Decrease in the stock out units	$1 - \sqrt{\left(1 - \frac{P}{100}\right)}$

Table 2.9 Comparison of % variation in the output parameters

P	$1-\sqrt{(1-\frac{P}{100})}$	$\sqrt{(1+\frac{P}{100})}-1$
10	0.051	0.049
20	0.105	0.095
30	0.163	0.140
40	0.225	0.183
50	0.293	0.225
60	0.367	0.265
70	0.452	0.304
80	0.553	0.342
90	0.684	0.378
100	1	0.414

Or

$$2 > \sqrt{\left(1+\frac{P}{100}\right)+\sqrt{\left(1-\frac{P}{100}\right)}}$$

Maximum value of right hand side (R.H.S.) corresponds to $P = 100$, therefore:

$$2 > \sqrt{\left(1+\frac{100}{100}\right)}+0$$

Or

$$2 > \sqrt{2}$$

And, that is true.

2.4 Demand Estimation for Variety of Products

While estimating the demand of products, few factors should be taken care of. These factors vary from product to product, pertaining to different types of business/industry. Some of these might be as follows:

(i) FMCG
(ii) Pharmaceutical
(iii) Agricultural
(iv) Hi-tech

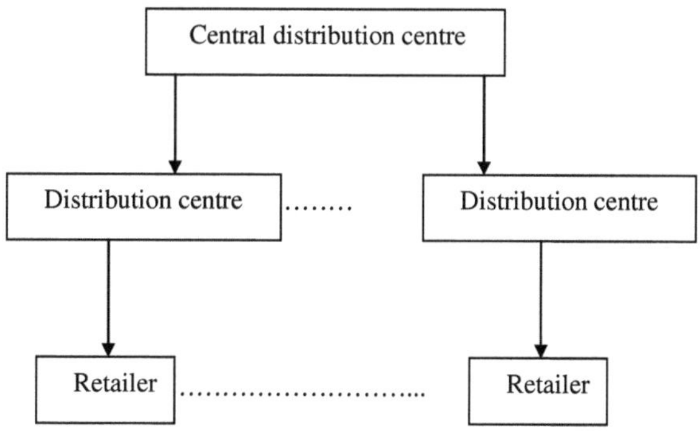

Fig. 2.6 Various levels in distribution system

2.4.1 *FMCG*

In case of the fast-moving consumer goods, demand might be estimated at various levels (Fig. 2.6) such as:

(a) Central distribution centre
(b) Distribution centre
(c) Retailer

Depending on the case, central distribution centre may ship the FMCG products to various distribution centres. Each distribution centre may also send the products to different retailers in a specific area. Estimation of a demand at distribution centres also depends on the transportation time and the duration for which the inventories are kept at a location before shipment. Additionally the following factors among others may also be considered:

(a) Season
(b) Geographical area
(c) Consumer group
(d) Retailer location
(e) Presence of the competitor products
(f) Price of the product
(g) Price discount offer by the company and its competitor
(h) Unavoidable price increase
(i) Advertisement of the product

2.4.2 *Pharmaceutical*

Additional attention may be paid to the following factors (Fig. 2.7) for an estimation of demand of the medicines:

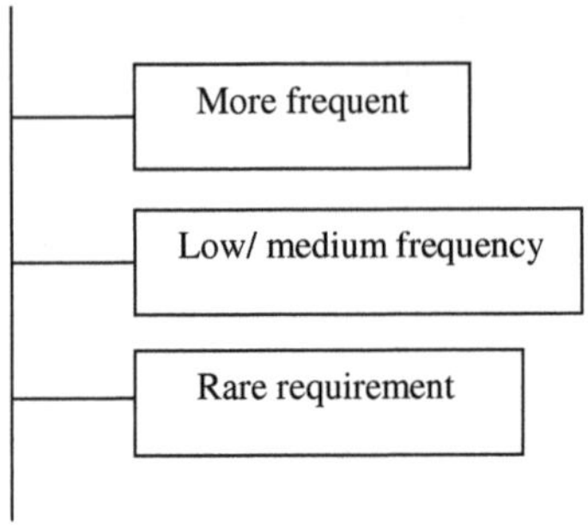

Fig. 2.7 Frequency of requirement for medicine

(a) Some of the medicines are more frequently required by the patients.
(b) There are requirements at low or medium frequency.
(c) Few medicines might be rarely required.

In view of the above, a judicious estimation regarding the demand should be made. Such estimation should also lead to the duration. Certain medicines might be needed more for a specific duration in a year. This may happen because of widespread disease in a particular season. This information is also helpful in determining the procurement ordering quantity on the basis of prediction.

In certain case, the retailer may also arrange for the medicines in a day or more if the customer is ready to wait. Such demands might be analyzed separately in the context of the following:

(i) Whether any stock should be kept for such occasional requirement?
(ii) If any stock is to be kept, what should be the quantity?

2.4.3 Agricultural

At a macro level, the following aspects would be helpful in understanding the agricultural scenario:

(a) Production of food grains
(b) Consumption side

In the context of the production of food grains, the following factors among others might be relevant:

(i) Area that is used for sowing a particular food grain
(ii) Yield per hectare
(iii) Growth rate concerning agriculture
(iv) Workers associated with agricultural tasks

(v) Area under irrigation

Concerning the consumption of food grains, the related factors among others might be:

(i) Earning level of the citizens
(ii) Population growth
(iii) Poverty level
(iv) Price of the food grains

These may help in macro level understanding of the supply-demand situation. In case where aggregate supply is more than the aggregate demand, then price reduction might happen affecting the demand positively. When aggregate demand is more than the aggregate supply, price rise happens and demand might decrease. However, after the farm output, the agricultural products including the food grains undergo several activities such as transportation, storage, and further processing among others.

In the context of fruits and vegetables, a demand should also be associated with the required freshness level. Qualitatively, the freshness level may be at lower level or higher level comparatively. The consumer who will purchase these items for consumption on the same day or for shorter period, might need comparatively lower freshness level. That is, their demand might be satisfied with less remaining shelf life also. On the other hand, the consumer who will purchase these items for consumption during the whole week or for longer periods might need comparatively higher freshness level. That is, their demands are satisfied with higher remaining shelf life.

2.4.4 Hi-Tech

In case of hi-tech products such as certain computer and telecommunication equipment, the product life cycle might be shorter. Thus, the demand estimation may also be linked to a shorter product life cycle. The existing version might become obsolete. Requirement of the exactly similar components/subassemblies also vanishes. Additionally, demands might be influenced by the following:

(i) Availability of substitute product:

In case where substitute product is available at the market place, the demands might be affected negatively.

(ii) Price:

When the price is on higher side, the demand might reduce. Relatively the demand might increase if price is on lower side.

(iii) Resale value:

In a situation, when resale value is considerable, it might be a considerable factor for some of the consumers for a decision regarding purchase.

(iv) Whether the citizens are technology savvy in general? or how much proportion is technology savvy?
(v) Purchasing power:

In general, higher purchasing power of the population may contribute toward purchase of the products.

(vi) Business requirement:

Because of a requirement of productivity improvement, hi-tech products may be used in the business process.

(vii) Peer/Societal pressure:

A section of the population may be inclined to purchase the product because of society pressure or peer group.

(viii) Govt. policy:

Government may also encourage the use of certain category of products in order to enhance its procedures/systems.

Chapter 3
Ordering Cost

While placing an item on order, few activities are to be undertaken by the concerned persons. These activities need to be analyzed for an estimation of the ordering cost.

3.1 Estimation of Ordering Cost

Difficulties in estimation of ordering cost increase in the following order for the case:

(i) Shopkeeper
(ii) Large retailer
(iii) Warehouse
(iv) Industry

In case of a shop, a helper may check the stock on shelf concerning an item and inform the owner/supervisor. Then the supervisor/owner spends some time in deciding whether this item should be ordered. There may be a need to do some paperwork related to placing an order. Alternatively, a telephone call might be enough if good relationship with the supplier exists.

For example, the helper takes 15 min to check the stock and to inform the superior. The superior takes 15 min to decide and do some paperwork for order placement. Daily wages of the helper are ₹200 for 10 h (say). Salary of the superior is proportionately ₹1000 for a day of 10 h (say).

Now, 15 min of the helper will cost:

$$\frac{15}{60}\times\frac{200}{10}=\frac{1}{4}\times 20=₹5$$

S. Sharma, *Inventory Parameters*, DOI 10.1007/978-981-10-4545-5_3

Similarly, 15 min of the superior will cost:

$$\frac{15}{60} \times \frac{1000}{10} = \frac{1}{4} \times 100 = ₹25$$

Therefore, the approximate ordering cost = 5 + 25 = ₹30.

In case of a large retailer, the following situation may arise:

(i) With the use of system, the concerned personnel might place an order for certain item after checking the stock.

(ii) The supplier company personnel may visit the retailer location and fill the shelf with the required item in the desirable quantities. However, the retail store employee will need to devote some time during their visit in order to execute the associated operations.

(iii) In case of the automated settings, system run time and associated expenditure need to be analyzed.

Activities/efforts associated with the ordering process should be listed and the concerned costs of resources are estimated in order to arrive at the ordering cost.

In case of a large warehouse, an order may be placed for the manufacturing plant. It is also possible that one order comprises of several items because the same plant produces all these items. Depending on the situation, the total ordering cost can be estimated for all such items. For each item, proportionately individual ordering cost may also be arrived at. Alternatively, an ordering cost may be associated with a group of items.

In case of an industrial organization, several/few supplier companies are selling the variety of components/raw materials to the buyer organization. With reference to the nature of operations, all or some of the following activities may be included in the ordering process for certain item:

(a) Before placing an order, the specifications of the component/item may need to be retrieved. The specifications may be revised/modified also.

(b) A tender may be invited for certain requirements. The tender might be global or applicable within the country. A publishing cost is incurred concerning the tender information after the efforts related to finalization of tender documents along with the terms and conditions. On receiving the proposals from the interested potential supplier companies, these are analyzed by the buyer company.
After selecting the suitable supplier, an order is placed by the buyer company. Relevant portion of the efforts and associated costs should be judiciously incorporated in the final estimation of the ordering cost for certain specific situation/business/item quantity desired by the buyer company.

(c) Before finalizing the order, the potential suppliers need to be evaluated on the basis of various details given by them including the past performance or other existing reputed buyer companies for their product. There might be a case when some of the suppliers have been doing business for the past considerable time with a buyer under consideration, i.e., the buyer company in the process of estimating the cost of ordering. In this situation, a potential supplier may also be evaluated on the basis of an available past record with the buyer concerning quality performance, after sales service, delivery performance, and price among other parameters. The supplier is finally selected before placing an order. The relevant efforts should be included in the ordering cost estimation.

Estimation of ordering cost for an industrial organization requires information concerning several factors. Figure 3.1 shows the ordered items used in the final product manufactured by the company. For example, an automobile company may order the tyres assembled into the car.

Ordering cost also depends on the factors such as:

(i) Market Scenario:
In case where it is buyer market, very less follow-up is needed after placing an order. Because of many suppliers, there is competition among them. After placing an order, the supplier tries to complete the order on time with an effort to satisfy the buyer company concerning other parameters. Less amount of chasing is needed by the buyer, and an indirect cost/hidden cost after placing an order is avoided.
On the other hand, if it is supplier market, a lot of efforts will go into follow-up activities by the buyer. Supplier companies in the market may be less in number. Supplier firm capacity may be lower than their total demands. Therefore, the buyer may not get the ordered quantities on time. In the real world, sometimes the buyer representative has to visit supplier premises to expedite their order at the supplier location. This results into the hidden cost after even placing the order.

(ii) Type of Product:
There are various types of products manufactured by the companies such as make to stock (MTS) and engineer to order (ETO) among others. Depending on the forecast, components/raw materials used in the final product may be ordered in MTS scenario. In contrast with the standard products concerning MTS, product design may be customized in ETO scenario. After getting the

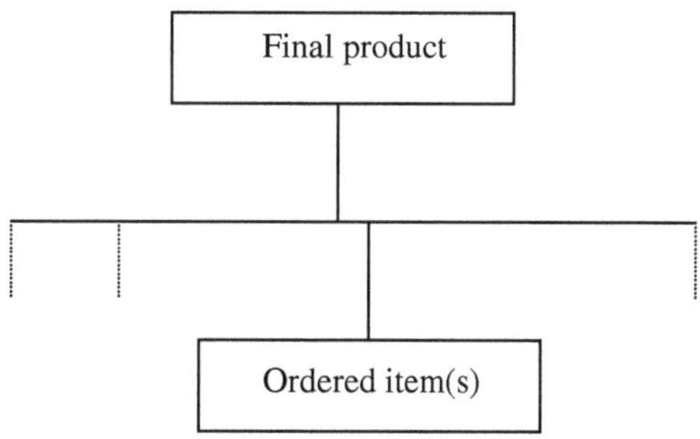

Fig. 3.1 Ordered item(s) by an industrial organization

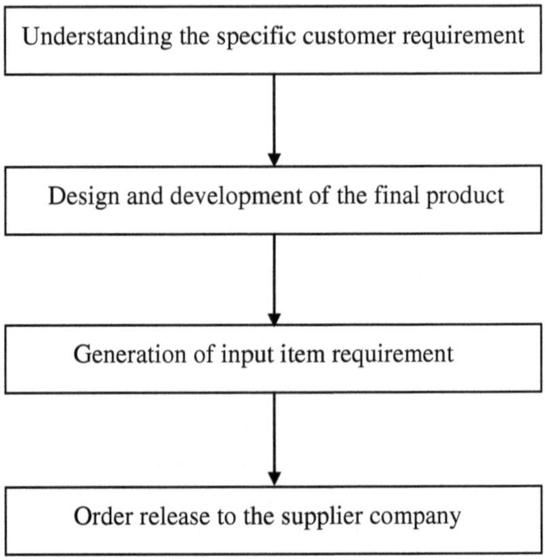

Fig. 3.2 Order release to the supplier in ETO scenario

precise customer requirement, product may be designed as shown in Fig. 3.2. As per the design and development of the final product, input item/components requirement in terms of quality/specifications/quantity is generated and order is released to the supplier company. The relevant efforts for order release to the supplier company should be listed and associated costs will be suitably incorporated into the estimation of ordering cost.

3.2 Ordering Cost Decrease

As the organization becomes matured/experienced, activities are more standardized and therefore a reduction in ordering cost is possible. Figure 3.3 shows various factors contributing toward a decrease in ordering cost.

Degree of automation may relate to:

(i) How many activities in the placement of order do not require human intervention or manual efforts at the buyer premises?
(ii) It is to be seen whether the related information sharing between buyer and supplier companies is smooth and through system with less human effort.
(iii) Whether appropriate degree of automation is chosen. This varies from a retailer to warehouse/an industrial organization.

In the context of precise stock status, it needs to be seen that an actual existing stock more or less is similar to what system says. If the system shows availability

Fig. 3.3 Contributing factors for ordering cost reduction

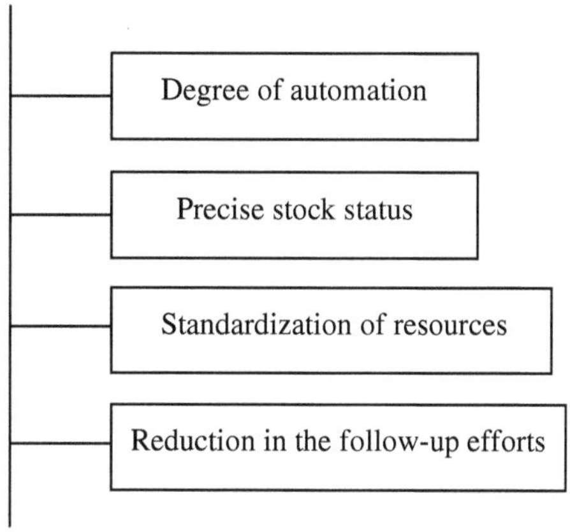

stock either higher or lower than the actual available quantity, then the ordering quantity will be erroneous.

Standardization of resources relates to:

(a) When the order is to be placed, specific person should take the initiative as far as possible.
(b) Specific person is authorized to make entries into the system concerning receipt or issue of the material/component.
(c) What steps are followed in the process of ordering and whether these are standard practices with certain standard time?

After placing an order, an effort is needed to follow up the order with the supplier in certain cases. If these efforts are tiresome and costly, a tendency is to order higher quantities in one order that may not be real optimum. Reduction in the follow-up efforts is necessary. This can be done by properly analyzing the supplier operations. Buyer company may also take the initiative in helping them so as to make their operations more efficient. If it is possible, supplier company personnel may be trained to make their activities more effective. Such joint efforts may be beneficial to both companies, i.e., supplier and buyer.

Now, let:

D = Annual demand
I = Annual inventory carrying cost per unit
C = Cost of ordering

$$\text{Optimum lot size } Q^* = \sqrt{\frac{2DC}{I}}$$

$$\text{Optimum cost } E^* = \sqrt{2DCI}$$

P = % decrease in the ordering cost

The decreased cost of ordering:

$$C_1 = \left(1 - \frac{P}{100}\right)C$$

$$\begin{aligned}\text{Decrease in the lot size} &= \sqrt{\frac{2DC}{I}} - \sqrt{\frac{2DC_1}{I}}\\ &= \sqrt{\frac{2DC}{I}} - \sqrt{\frac{2D(1-\frac{P}{100})C}{I}}\\ &= \sqrt{\frac{2DC}{I}}\left[1 - \sqrt{\left(1 - \frac{P}{100}\right)}\right]\end{aligned}$$

And:

$$\%\,\text{decrease in}\, Q = 1 - \sqrt{\left(1 - \frac{P}{100}\right)}$$

Now:

$$\begin{aligned}\text{Cost reduction} &= \sqrt{2DCI} - \sqrt{2DC_1I}\\ &= \sqrt{2DI}\left[\sqrt{C} - \sqrt{C_1}\right]\\ &= \sqrt{2DI}\left[\sqrt{C} - \sqrt{1 - \left(\frac{P}{100}\right)C}\right]\\ &= \sqrt{2DCI}\left[1 - \sqrt{1 - \left(\frac{P}{100}\right)}\right]\end{aligned}$$

And:

$$\%\,\text{decrease in}\, E = 1 - \sqrt{\left(1 - \frac{P}{100}\right)}$$

Table 3.1 Effects on parameters with respect to ordering cost decrease

% Decrease in *C*	5%	10%	15%	20%	25%	30%
C	28.5	27	25.5	24	22.5	21
Q	58	57	55	54	52	50
% Decrease in *Q*	2.5%	5.1%	7.8%	10.6%	13.4%	16.3%
E	585	569	553	537	520	502
% Decrease in *E*	2.5%	5.1%	7.8%	10.6%	13.4%	16.3%

Example 3.1
Consider the following values:

D	*C*	*I*	*Q*	*E*
600	30	10	60	600

Analyze whether there is a decrease in *C* in the following:

% Decrease in *C*	5%	10%	15%	20%	25%	30%
C	28.5	27	25.5	24	22.5	21

While decreasing the ordering cost, it is expected that the output parameters will decrease with the implementation of optimizing the scenario. These are shown in Table 3.1. Unlike the demand decrease, there is no chance of profit loss and it is a gain to the company. This gain is despite the lot size reduction and associated higher frequency of ordering.

3.3 Ordering Cost Increase

Although companies make efforts to decrease their ordering cost, yet there are certain situations in which increase in such cost is observed. Figure 3.4 shows various factors contributing toward an increase in ordering cost.

Salaries of the employees usually increase every year in the organization. Because of this reason, there is a chance of increase in ordering cost. While estimating such cost, an example is provided in Sect. 3.1. Helper time and supervisor time associated with the process of ordering were converted to cost using wages/salary. If these are increased, then upward variation in the ordering cost may not be avoided.

In a large organization, there may be necessary approvals at various levels before the release of the orders. In this way, total time in the process of ordering stretches contributing to the ordering cost increase.

In case of engineer to order (ETO) products, design/drawing is prepared by the engineering/design section. Frequent changes in the design might be due to one or more of the following reasons:

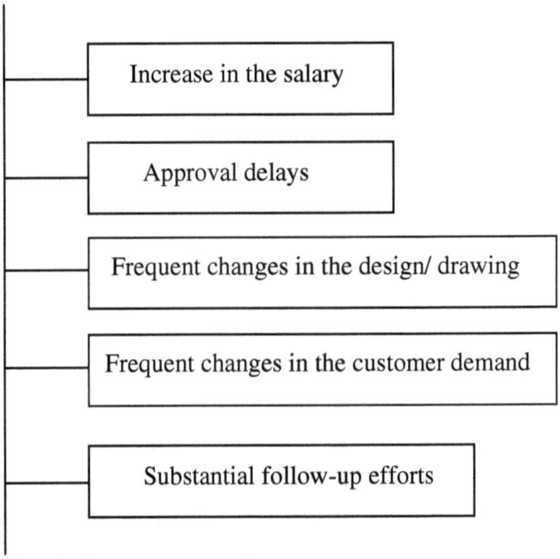

Fig. 3.4 Contributing factors for ordering cost increase

(i) Lack of consensus among the design team members.
(ii) Because of shortage of time, partial design may be forwarded to the purchase/production planning (so that they may at least start some of the work). Sometimes, there may be request for change in certain component design from the other departments/sections for ease in production/procurement. Therefore, the effort and time increases in the design.
(iii) Partial design was forwarded earlier and work already started for that remains. Meanwhile, request for change is received internally within the company. This may also lead to the revision of design that still remains and some efforts might have already gone waste.

Customers/customer companies of a manufacturing organization may seek change in their demands. Because of long-term relationships, such upward/downward change in customer demand may need to be entertained by the manufacturing company. However, this may require additional efforts in ordering for revised demands of components/raw materials that will be procured in order to make the finished product.

Substantial follow-up activities involving considerable efforts in order to get the desired quantities of an item from the supplier company indicate toward hidden cost in ordering. This factor also contributes toward ordering cost increase.

Now, let:

P = % increase in the ordering cost

Therefore, the increased cost of ordering:

$$C_1 = \left(1 + \frac{P}{100}\right) C$$

$$\text{Increase in the lot size} = \sqrt{\frac{2DC_1}{I}} - \sqrt{\frac{2DC}{I}}$$
$$= \sqrt{\frac{2D(1+\frac{P}{100})C}{I}} - \sqrt{\frac{2DC}{I}}$$
$$= \sqrt{\frac{2DC}{I}}\left[\sqrt{\left(1+\frac{P}{100}\right)} - 1\right]$$

And:

$$\%\ \text{increase in } Q = \sqrt{\left(1+\frac{P}{100}\right)} - 1$$

Now:

$$\text{Additional cost incurred} = \sqrt{2DC_1I} - \sqrt{2DCI}$$
$$= \sqrt{2DI}\left[\sqrt{C_1} - \sqrt{C}\right]$$
$$= \sqrt{2DI}\left[\sqrt{1+\left(\frac{P}{100}\right)C} - \sqrt{C}\right]$$
$$= \sqrt{2DCI}\left[\sqrt{1+\left(\frac{P}{100}\right)} - 1\right]$$

And:

$$\%\ \text{increase in } E = \sqrt{\left(1+\frac{P}{100}\right)} - 1$$

Example 3.2
Implement an ordering cost increase with reference to the following values:

D	C	I	Q	E
600	30	10	60	600

Where the percentage increase in ordering cost is as follows:

% Increase in C	5%	10%	15%	20%	25%	30%
C	31.5	33	34.5	36	37.5	39

Table 3.2 Effects on parameters with respect to ordering cost increase

% Increase in C	5%	10%	15%	20%	25%	30%
C	31.5	33	34.5	36	37.5	39
Q	61	63	64	66	67	68
% Increase in Q	2.5%	4.9%	7.2%	9.5%	11.8%	14.0%
E	615	629	643	657	671	684
% Increase in E	2.5%	4.9%	7.2%	9.5%	11.8%	14.0%

With reference to the increased ordering cost, the output parameters are shown in Table 3.2. These parameters, i.e., lot size and related cost increase.

3.4 Inclusion of Shortages

With an inclusion of shortages or stock outs, let the additional parameters be as follows:

J = Maximum units of shortages
K = Annual stock out cost per unit

Now, from Eqs. (1.1), (1.2), and (1.3), the output parameters are:

$$Q^* = \sqrt{\frac{2DC(K+I)}{IK}}$$

$$J^* = \sqrt{\frac{2DCI}{K(K+I)}}$$

$$E^* = \sqrt{\frac{2DCIK}{(K+I)}}$$

3.4.1 Reduction in Ordering Cost

With % reduction in ordering cost (P) and reduced ordering cost C_1:
Decrease in the lot size:

$$= \sqrt{\frac{2DC(K+I)}{IK}} - \sqrt{\frac{2D\left(1-\frac{P}{100}\right)C(K+I)}{IK}}$$

$$= \sqrt{\frac{2DC(K+I)}{IK}}\left[1 - \sqrt{\left(1-\frac{P}{100}\right)}\right]$$

$$\text{Decrease in the optimum stock out units} = \sqrt{\frac{2DCI}{K(K+I)}} - \sqrt{\frac{2DC_1I}{K(K+I)}}$$

$$= \sqrt{\frac{2DCI}{K(K+I)}} - \sqrt{\frac{2D\left(1-\frac{P}{100}\right)CI}{K(K+I)}}$$

$$= \sqrt{\frac{2DCI}{K(K+I)}}\left[1 - \sqrt{\left(1-\frac{P}{100}\right)}\right]$$

And, decrease in the total related cost:

$$= \sqrt{\frac{2DCIK}{(K+I)}} - \sqrt{\frac{2DC_1IK}{(K+I)}}$$

$$= \sqrt{\frac{2DCIK}{(K+I)}}\left[1 - \sqrt{\left(1-\frac{P}{100}\right)}\right]$$

Example 3.3
Consider the following information.

D	*C*	*I*	*K*	*Q*	*E*	*J*
600	30	10	200	61.48	585.54	2.93

Implement the ordering cost reduction as follows:

% Decrease in *C*	5%	10%	15%	20%	25%	30%
C	28.5	27	25.5	24	22.5	21

The decreased output parameters are shown in Table 3.3. With the inclusion of stock outs, lot size has increased. This is in order to handle the shortages in an optimal manner. This increase is because of multiplication of the following factor:

Table 3.3 Effects on parameters with stock out corresponding to ordering cost decrease

% Decrease in *C*	5%	10%	15%	20%	25%	30%
C	28.5	27	25.5	24	22.5	21
Q	59.92	58.33	56.68	54.99	53.24	51.44
% Decrease in *Q*	2.53%	5.13%	7.80%	10.56%	13.40%	16.33%
J	2.85	2.78	2.70	2.62	2.54	2.45
% Decrease in *J*	2.53%	5.13%	7.80%	10.56%	13.40%	16.33%
E	570.71	555.49	539.84	523.72	507.09	489.90
% Decrease in *E*	2.53%	5.13%	7.80%	10.56%	13.40%	16.33%

$$\sqrt{\frac{(K+I)}{K}}$$

As $(K + I) > K$, this factor is always greater than 1.

However, the related total costs decrease in comparison with the situation when stock outs are not allowed. This decrease is because of the multiplication of the following factor:

$$\sqrt{\frac{K}{(K+I)}}$$

As $K < (K + I)$, this factor is always less than 1.

As shown in Fig. 3.5, it is possible to reduce the ordering cost initially by way of eliminating the inefficiencies. This may happen with the existing resources also and no substantial additional investment is needed. However, beyond certain level of reduced ordering cost (shown at the lowest level of such cost), it is difficult without certain additional investment. Such investment may relate to the following:

(a) An improved IT tool
(b) Training of the persons responsible for ordering process
(c) Better information system for internal as well as external communication, i.e., with the supplier companies
(d) Coordination effort including visit of the supplier company
(e) Training of the relevant supplier company personnel

In most of the cases, such an investment incorporates time, or cost, or both (as time can easily be converted to cost). These efforts contribute toward successful

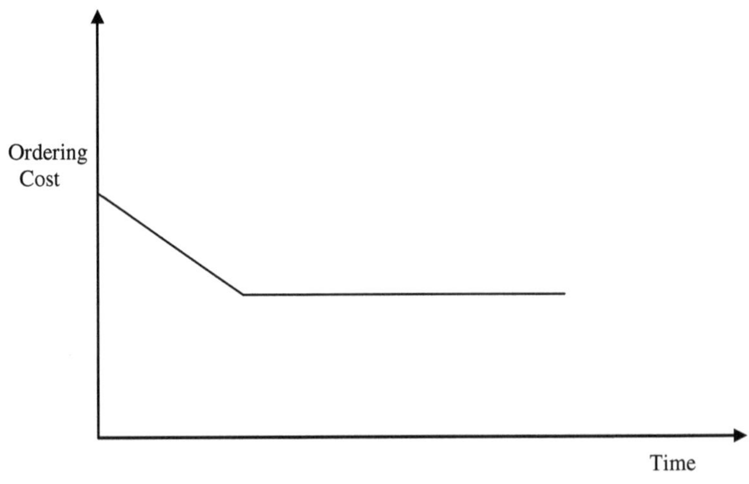

Fig. 3.5 Reducing the cost of ordering

long-term/strategic relationship with the supplier, and ordering cost (also including hidden cost associated with that) can be reduced to the fullest extent.

3.4.2 *Increase in Ordering Cost*

As an increased ordering cost $C_1 = \left(1 + \frac{P}{100}\right)C$, and lot size $Q^* = \sqrt{\frac{2DC(K+I)}{IK}}$, increase in lot size:

$$= \sqrt{\frac{2D\left(1 + \frac{P}{100}\right)C(K+I)}{IK}} - \sqrt{\frac{2DC(K+I)}{IK}}$$

$$= \sqrt{\frac{2DC(K+I)}{IK}}\left[\sqrt{\left(1 + \frac{P}{100}\right)} - 1\right]$$

$$\text{Increase in the optimum stock out units} = \sqrt{\frac{2DC_1I}{K(K+I)}} - \sqrt{\frac{2DCI}{K(K+I)}}$$

$$= \sqrt{\frac{2D\left(1 + \frac{P}{100}\right)CI}{K(K+I)}} - \sqrt{\frac{2DCI}{K(K+I)}}$$

$$= \sqrt{\frac{2DCI}{K(K+I)}}\left[\sqrt{\left(1 + \frac{P}{100}\right)} - 1\right]$$

And, increase in the total related cost:

$$= \sqrt{\frac{2DC_1IK}{(K+I)}} - \sqrt{\frac{2DCIK}{(K+I)}}$$

$$= \sqrt{\frac{2DCIK}{(K+I)}}\left[\sqrt{\left(1 + \frac{P}{100}\right)} - 1\right]$$

Example 3.4
Consider the following information:

D	C	I	K	Q	E	J
600	30	10	200	61.48	585.54	2.93

Table 3.4 Effects on parameters with stock out corresponding to ordering cost increase

% Increase in C	5%	10%	15%	20%	25%	30%
C	31.5	33	34.5	36	37.5	39
Q	63	64.5	65.93	67.35	68.74	70
% Increase in Q	2.5%	4.9%	7.2%	9.5%	11.8%	13.9%
E	600	614.12	628	641.4	654.65	667.6
% Increase in E	2.5%	4.9%	7.3%	9.5%	11.8%	14.0%
J	3	3.07	3.14	3.21	3.27	3.33
% Increase in J	2.4%	4.8%	7.2%	9.6%	11.6%	13.7%

Implement the ordering cost increase as follows:

% Increase in C	5%	10%	15%	20%	25%	30%
C	31.5	33	34.5	36	37.5	39

After incorporating the stock out situation and an increase in cost of ordering, the effects on parameters are represented in Table 3.4.

3.5 Interaction of Ordering Cost and Demand

With respect to an increase in demand (i.e., from D to D_1), procurement lot size also increases in addition to total cost increase. This is represented by Fig. 3.6.

Although profitability increases by way of selling larger quantity of a product, however in order to enhance the level of profit furthermore, the wholesaler/retailer might be interested in keeping the total cost level similar to previous scenario (i.e., corresponding to demand level D). This is possible by a reduction in the ordering cost among other options. Such interaction between ordering cost and demand is useful for inventory parameters management.

Example 3.5
The existing data are as follows:

Annual demand, $D = 600$ units
Ordering cost, $C = ₹30$
Annual inventory carrying cost per unit, $I = ₹10$

And the total related annual cost from Eq. (1.5).

$$\begin{aligned} E &= \sqrt{2DCI} \\ &= \sqrt{2 \times 600 \times 30 \times 10} \\ &= ₹600 \end{aligned}$$

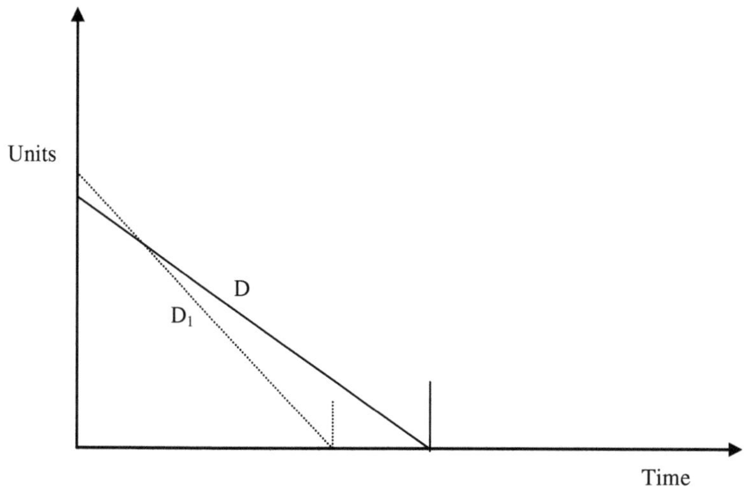

Fig. 3.6 Increase in demand

Now, consider the demand increase by 10%, i.e., D_1 = 660 and the associated cost:

$$E = ₹629.29$$

With the increased demand also, if the objective is to maintain a previous total cost level, then the desired ordering cost, C_1, can be obtained as follows:

$$\sqrt{2 \times 660 \times C_1 \times 10} = 600$$

Or

$$2 \times 660 \times C_1 \times 10 = 600 \times 600$$

Or

$$C_1 = \frac{600 \times 600}{2 \times 660 \times 10}$$

Or

$$C_1 = ₹27.27$$

% reduction in the ordering cost is approximately 9.09%, i.e., less than the demand increase by 10%. In order to generalize, the notation is as follows:

P = % increase in demand
M = % reduction in the ordering cost

Now:

$$\sqrt{2D_1C_1I} = \sqrt{2DCI}$$

Or

$$2D_1C_1I = 2DCI$$

Or

$$D_1C_1 = DC$$

Or

$$\left(1 + \frac{P}{100}\right)D\left(1 - \frac{M}{100}\right)C = DC$$

Or

$$\left(1 + \frac{P}{100}\right)\left(1 - \frac{M}{100}\right) = 1$$

Or

$$\left(1 - \frac{M}{100}\right) = \frac{1}{(1 + P/100)}$$

Or

$$\frac{M}{100} = 1 - \frac{1}{(1 + P/100)}$$

Or

$$\frac{M}{100} = \frac{(P/100)}{(1 + P/100)}$$

Or

$$M = \frac{P}{(1 + P/100)}$$

In order to know the variation in M (Table 3.5) with respect to P, the derived expression can be used.

Values of M are lower than the respective values of P. Furthermore, M is less sensitive with respect to the higher values of P.

Table 3.5 Variation in M (ordering cost) with respect to P (demand)

S. No.	P	$M = \frac{P}{(1+P/100)}$
1	10	9.09
2	20	16.67
3	30	23.08
4	40	28.57
5	50	33.33

When change is triggered by the reduction in ordering cost, then the management response might be in the form of exploring a potential increase in demand. For example, consider 10% reduction in ordering cost. Now:

$$\sqrt{2 \times D_1 \times 27 \times 10} = 600$$

Or

$$2 \times D_1 \times 27 \times 10 = 600 \times 600$$

Or

$$D_1 = \frac{600 \times 600}{2 \times 27 \times 10}$$

Or

$$D_1 = 666.67$$

% variation in demand is more than the % variation in ordering cost. In order to generalize, let:

P = % reduction in ordering cost
M = % increase in the potential demand

Now, for similar total cost level:

$$\sqrt{2D_1C_1I} = \sqrt{2DCI}$$

Or

$$D_1C_1 = DC$$

Or

$$\left(1 + \frac{M}{100}\right)D\left(1 - \frac{P}{100}\right)C = DC$$

Or

$$\left(1+\frac{M}{100}\right)\left(1-\frac{P}{100}\right)=1$$

Or

$$\left(1+\frac{M}{100}\right)=\frac{1}{(1-P/100)}$$

Or

$$\frac{M}{100}=\frac{1}{(1-P/100)}-1$$

Or

$$\frac{M}{100}=\frac{(P/100)}{(1-P/100)}$$

Or

$$M=\frac{P}{(1-P/100)}$$

In order to know the variation in M (Table 3.6) with respect to P, the derived expression can be used.

Values of M are higher than the respective values of P. Furthermore, M is more sensitive with respect to the higher values of P.

Comparing with Table 3.5, values of M are higher. This is because:

$$\frac{P}{(1-P/100)}>\frac{P}{(1+P/100)}$$

Or

$$(1-P/100)<(1+P/100)$$

Table 3.6 Variation in M (demand) with respect to P (ordering cost)

S. No.	P	$M=\frac{P}{(1-P/100)}$
1	10	11.11
2	20	25.00
3	30	42.86
4	40	66.67
5	50	100

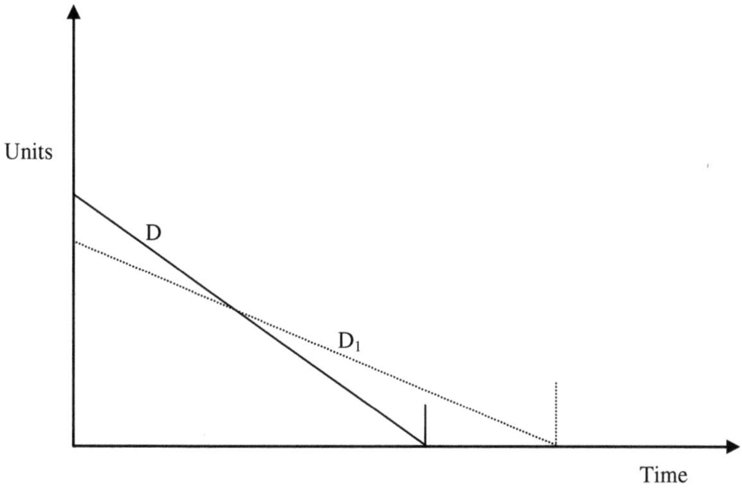

Fig. 3.7 Decrease in demand

And that is true.

Because of the demand reduction, procurement lot size decreases (Fig. 3.7) and also total cost reduces.

Example 3.6

In case of demand reduction, there is profit loss because of lower revenue. In order to compensate for this, the ordering cost may be reduced. With reference to the previous example, if demand decreases by 10%, then:

$$D_1 = 540$$

And total cost = ₹569.

Although cost is reduced, however in addition to this, the retailer may like to offset the loss of profit also. Assume that the loss of profit is at the rate of ₹1 per product. Therefore the reduction in profit:

$$1 \times 60 = ₹60$$

The accounting procedure may differ from company to company. However, it is always beneficial to keep the target cost as low/feasible as possible. Assume that it is now:

$$569 - 60 = ₹509$$

The desired lower level of ordering cost can be obtained as follows:

$$\sqrt{2 \times 540 \times C_1 \times 10} = 509$$

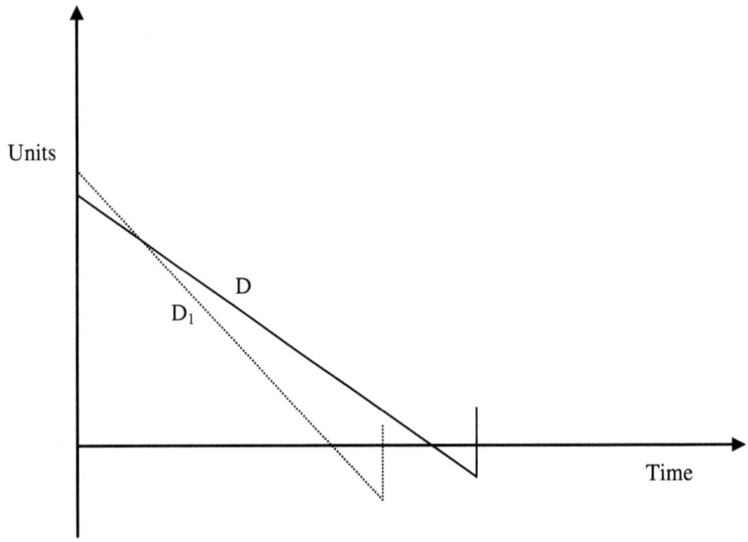

Fig. 3.8 Increase in demand (with stock out)

Or

$$C_1 = ₹23.99$$

Inclusion of stock out:
While incorporating the stock out, the case is represented by Fig. 3.8 when demand is increased from D to D_1.

Example 3.7
Consider the following information:

D	C	I	K	Q	E	J
600	30	10	200	61.48	585.54	2.93

With the increased demand (by 15%) also, if the objective is to maintain a previous total cost level, then the desired ordering cost, C_1, can be obtained as follows:

$$\sqrt{\frac{2 \times 690 \times C_1 \times 10 \times 200}{(200 + 10)}} = 585.54$$

Or

$$2 \times 690 \times C_1 \times 10 \times 200 = 585.54 \times 585.54 \times 210$$

Or

$$C_1 = \frac{585.54 \times 585.54 \times 210}{2 \times 690 \times 10 \times 200}$$

Or

$$C_1 = ₹26.09$$

In case where the change is initiated with a reduction in ordering cost (say, by 15%), then the potential increased demand in order to maintain a similar total cost is evaluated as:

$$\sqrt{\frac{2 \times D_1 \times 25.5 \times 10 \times 200}{(200 + 10)}} = 585.54$$

Or

$$2 \times D_1 \times 25.5 \times 10 \times 200 = 585.54 \times 585.54 \times 210$$

Or

$$D_1 = \frac{585.54 \times 585.54 \times 210}{2 \times 25.5 \times 10 \times 200}$$

Or

$$D_1 = 705.88$$

While incorporating the stock out, the case is represented by Fig. 3.9 when demand is reduced from D to D_1.

Example 3.8
In case of demand reduction, there is profit loss because of lower revenue. In order to compensate for this, the ordering cost may be reduced. With reference to the previous example, if demand decreases by 15%, then:

$$D_1 = 510$$

And total cost = ₹539.84.

Although cost is reduced, however in addition to this, the retailer may like to offset the loss of profit also. Assume that the loss of profit is at the rate of ₹1 per product. Therefore the reduction in profit:

$$1 \times 90 = ₹90$$

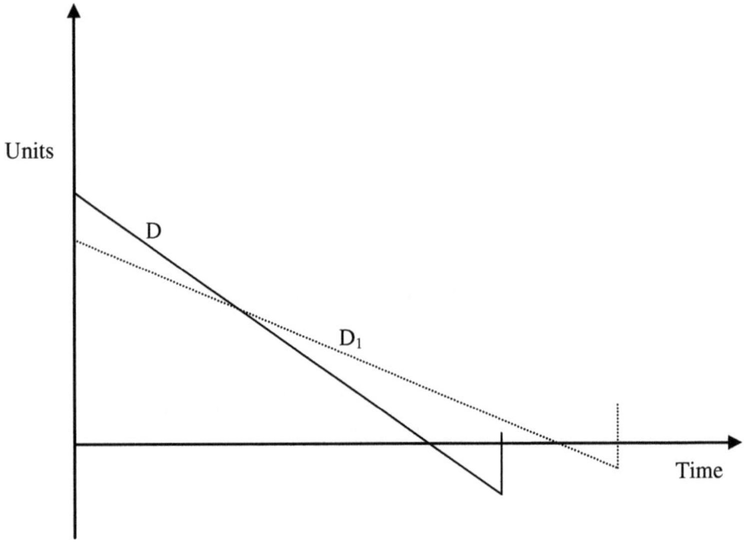

Fig. 3.9 Decrease in demand (with stock out)

The accounting procedure may differ from company to company. However, it is always beneficial to keep the target cost as low/feasible as possible. Assume that it is now:

$$539.84 - 90 = ₹449.84$$

The desired lower level of ordering cost can be obtained as follows:

$$\sqrt{\frac{2 \times 510 \times C_1 \times 10 \times 200}{(200 + 10)}} = 449.84$$

Or

$$2 \times 510 \times C_1 \times 10 \times 200 = 449.84 \times 449.84 \times 210$$

Or

$$C_1 = \frac{449.84 \times 449.84 \times 210}{2 \times 510 \times 10 \times 200}$$

Or

$$C_1 = ₹20.38$$

3.6 Ordering Cost Estimation for Variety of Products

Ordering cost depends on the relevant procedures and resources involved among other aspects. This varies according to the type of industry or product such as:

(i) FMCG
(ii) Pharmaceutical
(iii) Food products
(iv) Hi-tech

3.6.1 *FMCG*

In case of the fast-moving consumer goods, frequent replenishment is necessary. An estimation of the ordering cost will vary from level to level and also considering the degree of automation. The following situations might be relevant:

(a) In case of a shopkeeper/small retailer, degree of automation may be very low. Depending on their experience and judgment, they may place order for the desired quantities to the distributor/wholesaler. In a small town or village, shopkeeper might visit the nearest city to bring the desired goods every week. In principle, the efforts needed for placing an order should be accounted in terms of money value in order to arrive at an appropriate ordering cost.
(b) In case of a large retailer, degree of automation may be high. As soon as the stock decreases to certain level, an order may be placed. However, the overheads and managerial control needed in the process should be accounted for efforts and related cost in order to arrive at suitable ordering cost for furthermore analysis. In certain cases, the FMCG company representatives take the responsibility to replenish the goods at the large retailer location. However, the coordination efforts in the context of ordering process should be carefully examined and the ordering cost might be estimated wherever relevant.

3.6.2 *Pharmaceutical*

In case of the sale of medicines, the consumers might be individuals or the hospital may purchase for onward distribution to patients. The pharmaceutical company representative or distributor representative may visit the retailer periodically. With the help of online placement of order, the representative might place an order for the desired quantities of medicines. The related efforts may be quantified to estimate an ordering cost. Else the retailer may keep a track of the stock and place an order for suitable quantities. Quantities in each order may be in accordance with the demand estimate among other factors.

In case of the pharmaceutical equipment sale, it may not be necessarily available at the retail store. Depending on the occasional requirement of the customer, retailer may place specific order when the customer is ready to wait. However, special efforts need to be made for such order and its expediting in case of delay. Ordering cost estimate may be in correspondence with special effort and time.

3.6.3 Food Products

Food products/agricultural output may be of wide variety such as:

(a) Fruits
(b) Vegetables
(c) Food grains and flour
(d) Output of the food processing companies

A retailer will place an order for the wholesaler. Similarly, the wholesaler may place the order for furthermore upstream agencies in the chain. As shown in Fig. 3.10, cost of placing an order may depend on few factors in the present context.

Ordering cost estimate will vary as per the following factors:

(i) Ordering system adopted
(ii) Resources involved in the process of ordering
(iii) Degree of expediting each order in case of delay/uncertainty
(iv) Scale of operations, i.e., small/medium/large retailer/company

Fig. 3.10 Factors affecting ordering cost estimate

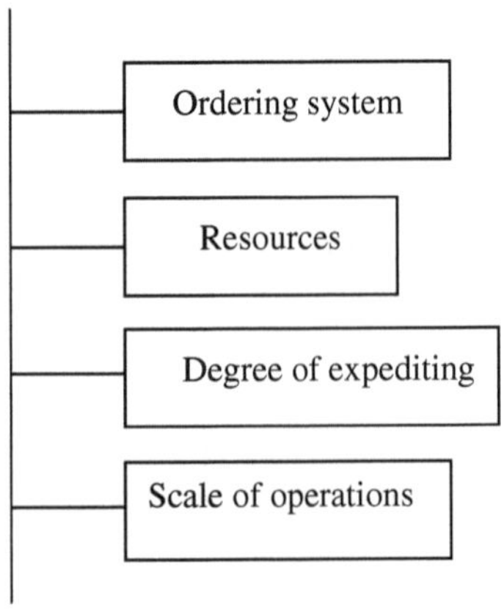

3.6.4 Hi-Tech

Hi-tech products might include among others:

(a) Computing equipment/computer system
(b) Telecommunication equipment

In case of certain critical component/circuit, number of suppliers may be limited. Furthermore, if the supplier capacity is limited, there is a lot of dependence on the supplier company. In case where a company is open for supply to a competitor also, order expediting might take a lot of efforts and time. Such aspects may be additionally incorporated into the estimation of ordering cost. Because of shorter product life cycle, a change in the supplier also may not be strategically advised.

[illegible]

Chapter 4
Inventory Carrying Cost

Certain costs are involved while carrying the inventories at the retailer location, warehouse, and stores of industrial organizations. Variation in the inventory carrying or holding costs needs to be analyzed with or without back orders. For this purpose, precise estimation of carrying cost is very necessary.

4.1 Carrying Cost Estimation

The annual inventory carrying cost per unit is estimated after suitable consideration of the following aspects:

(a) While keeping certain inventory item in store, space is occupied by it. The space is self-owned or hired. In case of the hired space, rent is paid. When the space is owned by the retailer or company and it is fully utilized, there is loss of opportunity to keep other item.

(b) Certain level of investment is needed in order to procure the inventories. Some other benefits could be obtained if procurement of inventories is not there, such as an earned interest and benefit associated with investment elsewhere among others.

(c) If evaporation losses are there while keeping certain liquid inventory items for longer time, then such losses should also be accounted. These are also dependent on the quantity available in a container among other factors. While handling the items also, some losses may take place.

(d) Certain type of inventory items may have shelf life such as food items and medicines, i.e., these expire after specified time. In case where such items are procured in larger quantities and stored for longer time, some portion of quantity may remain unused and, meanwhile, shelf life would be over resulting into loss.

S. Sharma, *Inventory Parameters*, DOI 10.1007/978-981-10-4545-5_4

(e) While storing some items, there is specific temperature requirement. Refrigerated environment may be needed. Suitable air-conditioning with specific humidity requirement might be necessary. These efforts involve cost contributing toward the inventory carrying.
(f) Space required for storing the items may need regular cleaning, proper illumination, suitable ventilation, and frequent supervision. Some methods for preserving specific type of items are also necessary.

These aspects are also summarized in Fig. 4.1. For a specific item, relevant factors contributing to inventory carrying cost should be identified and then carrying or holding cost is estimated per unit item for specified period. In some cases, smaller items are put in one container and it is easier to estimate for one container. In case of the need, proportionately, such costs per unit item may easily be obtained. Similarly, it may be more convenient to estimate for a lot of few hundred items (say) and annual inventory holding cost may be apportioned per unit item. Supervision cost including human effort and equipment operational cost might be obtained for few hundred or thousands of items in a location unit before arriving at the carrying cost for a judicious unit of items depending on relevant factors. Equipment operational cost might refer to refrigerator/air conditioner running costs for certain space or for quantities of items that are in the form of say solid or liquid.

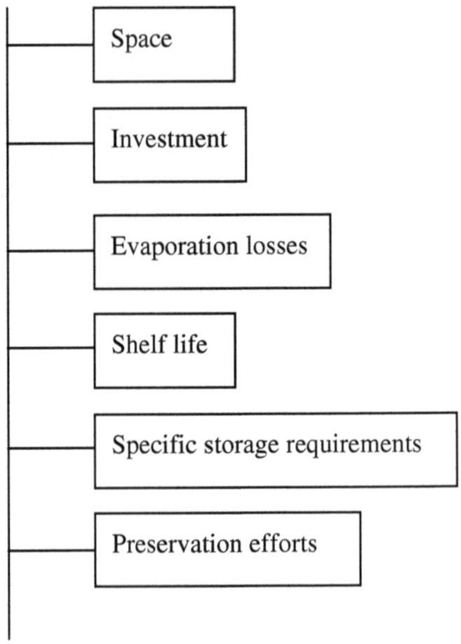

Fig. 4.1 Some aspects related to carrying cost estimation

4.2 Increase in Holding Cost

An increase in the inventory holding or carrying cost might happen because of the following reasons:

(i) There are other better opportunities available now, and the cost of lost opportunity is obviously higher.
(ii) Because of the changed economic scenario and policies, cost of capital might be relatively higher. If the cost of capital is relevant for a business in the context of procurement, then it may affect inventory holding cost.
(iii) Because of power crisis, it may be difficult to maintain the required temperature continuously. Spoilage of the products may be increasing due to this reason.
(iv) Because of salary/wage increase, supervision cost attributed to inventory holding might be higher.
(v) Cost of hiring space increases and it contributes to a higher carrying cost if it is relevant.
(vi) Earlier, the supplier of an input item was ready to dispatch the desired quantity of that item on credit payment. But now, that supplier stopped credit facility for certain reasons. This also increases the carrying or holding cost.
(vii) Credit period might be shortened now, and, therefore, the holding cost increases relatively.

It helps to quantify the effects of an increase in holding cost on the parameters such as procurement lot size and related cost.

Example 4.1
Consider the following input data:

Annual demand, $D = 600$ units
Ordering cost, $C =$ ₹30
Annual inventory carrying cost per unit, $I =$ ₹10

As the optimal output parameters are given by:

$$Q^* = \sqrt{\frac{2DC}{I}} = 60 \text{ units}$$

$$E^* = \sqrt{2DCI} = ₹600$$

Implement the following increase in inventory holding cost and comment:

% Increase in I	5%	10%	15%	20%	25%	30%
I	10.5	11	11.5	12	12.5	13

Because of an increase in holding cost, lot size decreases and related cost increases. These effects (i.e., approximate figures concerning absolute values as well as percentage variation) are shown in Table 4.1.

Table 4.1 Effects on parameters with respect to increase in holding cost

% Increase in I	5%	10%	15%	20%	25%	30%
I	10.5	11	11.5	12	12.5	13
Q	59	57	56	55	54	53
% Decrease in Q	2.4%	4.7%	6.7%	8.7%	10.6%	12.3%
E	615	629	643	657	671	684
% Increase in E	2.5%	4.9%	7.2%	9.5%	11.8%	14.0%

Let:

P = % variation in parameter

For the present discussion, P refers to the % increase in inventory carrying or holding cost, therefore the increased holding cost:

$$I_1 = \left(1 + \frac{P}{100}\right) I$$

$$\begin{aligned}
\text{Decrease in the lot size} &= \sqrt{\frac{2DC}{I}} - \sqrt{\frac{2DC}{I_1}} \\
&= \sqrt{\frac{2DC}{I}} - \sqrt{\frac{2DC}{\left(1 + \frac{P}{100}\right) I}} \\
&= \sqrt{\frac{2DC}{I}} \left[1 - \sqrt{\frac{1}{\left(1 + \frac{P}{100}\right)}}\right]
\end{aligned}$$

And:

$$\% \text{ decrease in } Q = 1 - \sqrt{\frac{1}{\left(1 + \frac{P}{100}\right)}}$$

Now:

$$\begin{aligned}
\text{Additional cost incurred} &= \sqrt{2DCI_1} - \sqrt{2DCI} \\
&= \sqrt{2DC}\left[\sqrt{I_1} - \sqrt{I}\right] \\
&= \sqrt{2DC}\left[\sqrt{\left(1 + \frac{P}{100}\right) I} - \sqrt{I}\right] \\
&= \sqrt{2DCI}\left[\sqrt{\left(1 + \frac{P}{100}\right)} - 1\right]
\end{aligned}$$

And:

$$\%\,\text{increase in}\,E=\sqrt{\left(1+\frac{P}{100}\right)}-1$$

The obtained generalized results are summarized in Table 4.2.

Such analysis helps in knowing:

(i) The decrease/% decrease in optimum procurement batch size and subsequent release of storage space for alternative use.
(ii) Additional related cost and planning for its arrangement

Variation in cost is higher than that in the lot size. This is because:

$$\sqrt{\left(1+\frac{P}{100}\right)}-1>1-\sqrt{\frac{1}{\left(1+\frac{P}{100}\right)}}$$

Or

$$2<\sqrt{\left(1+\frac{P}{100}\right)}+\sqrt{\frac{1}{\left(1+\frac{P}{100}\right)}}$$

Or

$$4<\left[\sqrt{\left(1+\frac{P}{100}\right)}+\sqrt{\frac{1}{\left(1+\frac{P}{100}\right)}}\right]^2$$

Or

$$4<1+\frac{P}{100}+\frac{1}{1+(P/100)}+2$$

Table 4.2 Results with reference to % increase in holding cost

Decrease in the lot size	$\sqrt{\frac{2DC}{I}}\left[1-\sqrt{\frac{1}{\left(1+\frac{P}{100}\right)}}\right]$
% Decrease in the lot size	$1-\sqrt{\frac{1}{\left(1+\frac{P}{100}\right)}}$
Additional related cost	$\sqrt{2DCI}\left[\sqrt{\left(1+\frac{P}{100}\right)}-1\right]$
% Increase in cost	$\sqrt{\left(1+\frac{P}{100}\right)}-1$

Or

$$4 < 3 + \frac{P}{100} + \frac{1}{1 + (P/100)}$$

Or

$$4 < 3 + \frac{P}{100} + \frac{1}{1 + (P/100)}$$

Or

$$4 < 3 + \frac{(P/100) + (P/100)^2 + 1}{1 + (P/100)}$$

Or

$$4 < 3 + 1 + \frac{(P/100)^2}{1 + (P/100)}$$

Or

$$4 < 4 + \frac{(P/100)^2}{1 + (P/100)}$$

And that is true.

4.3 Decrease in Holding Cost

A decrease in holding cost might happen because of the following reasons:

(i) Enough suitable opportunities do not exist, and, therefore, the cost of lost opportunity is relatively lower.

(ii) Because of certain changes in overall economic scenario, cost of capital is comparatively lower. The inventory holding cost may be lower if the cost of capital is relevant in the context of purchase of inventories.

(iii) Because of better maintenance of required temperature continuously on account of uninterrupted power supply, spoilage of the product is relatively less.

(iv) With the use of technology, man power productivity is improved, and, therefore, the supervision cost is relatively lower.
(v) Because of an increasing competition/availability of storage space, cost of hiring space has gone down. If the company needs a hired space to keep the inventory items, then the holding cost might be lower.
(vi) Large number of suppliers of input items now exist and they are ready to offer the credit facility due to competition and also to become the preferred supplier company.
(vii) Credit period might be elongated, and, therefore, the holding cost decreases comparatively.

Parameters such as procurement lot size and related cost get affected by the decrease in holding cost, and it is of interest to analyze those.

Example 4.2
Consider the input data of Example 4.1. Analyze the decrease in holding cost as follows:

% Decrease in I	5%	10%	15%	20%	25%	30%
I	9.5	9	8.5	8	7.5	7

With the decrease in I, larger lot size can be procured with an overall decrease in the related cost. The results are shown in Table 4.3. These are useful in knowing:

(a) How much is an increase in procurement quantity and the related additional requirement for the storage space?
(b) How much is the reduction in relevant cost?

In order to generalize:

P = % decrease in inventory carrying or holding cost, therefore the decreased holding cost:

$$I_1 = \left(1 - \frac{P}{100}\right)I$$

Table 4.3 Effects on parameters with respect to decrease in holding cost

% Decrease in I	5%	10%	15%	20%	25%	30%
I	9.5	9	8.5	8	7.5	7
Q	62	63	65	67	69	72
% Increase in Q	2.6%	5.4%	8.5%	11.8%	15.5%	19.5%
E	585	569	553	537	520	502
% Decrease in E	2.5%	5.1%	7.8%	10.6%	13.4%	16.3%

$$\text{Increase in the lot size} = \sqrt{\frac{2DC}{I_1}} - \sqrt{\frac{2DC}{I}}$$
$$= \sqrt{\frac{2DC}{(1-\frac{P}{100})I}} - \sqrt{\frac{2DC}{I}}$$
$$= \sqrt{\frac{2DC}{I}}\left[\sqrt{\frac{1}{(1-\frac{P}{100})}} - 1\right]$$

And:

$$\%\,\text{increase in } Q = \sqrt{\frac{1}{(1-\frac{P}{100})}} - 1$$

Now:

$$\text{Cost reduction} = \sqrt{2DCI} - \sqrt{2DCI_1}$$
$$= \sqrt{2DC}\left[\sqrt{I} - \sqrt{I_1}\right]$$
$$= \sqrt{2DC}\left[\sqrt{I} - \sqrt{\left(1-\frac{P}{100}\right)I}\right]$$
$$= \sqrt{2DCI}\left[1 - \sqrt{\left(1-\frac{P}{100}\right)}\right]$$

And:

$$\%\,\text{decrease in } E = 1 - \sqrt{\left(1-\frac{P}{100}\right)}$$

These generalized results are also summarized in Table 4.4.

Table 4.4 Results with reference to % decrease in holding cost

Increase in the lot size	$\sqrt{\frac{2DC}{I}}\left[\sqrt{\frac{1}{(1-\frac{P}{100})}} - 1\right]$
% Increase in the lot size	$\sqrt{\frac{1}{(1-\frac{P}{100})}} - 1$
Cost reduction	$\sqrt{2DCI}\left[1 - \sqrt{(1-\frac{P}{100})}\right]$
% Decrease in cost	$1 - \sqrt{(1-\frac{P}{100})}$

In this case, the variation in lot size is higher than that in the cost. This is because:

$$\sqrt{\frac{1}{\left(1-\frac{P}{100}\right)}}-1>1-\sqrt{\left(1-\frac{P}{100}\right)}$$

Or

$$2<\sqrt{\frac{1}{\left(1-\frac{P}{100}\right)}}+\sqrt{\left(1-\frac{P}{100}\right)}$$

Or

$$4<\frac{1}{1-(P/100)}+1-\frac{P}{100}+2$$

Or

$$4<3+\frac{1}{1-(P/100)}-\frac{P}{100}$$

Or

$$4<3+\frac{1-(P/100)(1-P/100)}{1-(P/100)}$$

Or

$$4<3+\frac{1-(P/100)+(P/100)^2}{1-(P/100)}$$

Or

$$4<3+1+\frac{(P/100)^2}{1-(P/100)}$$

Or

$$4<4+\frac{(P/100)^2}{1-(P/100)}$$

And that is true for all practical values of $P < 100$.

4.4 Backordering Scenario

When the shortages/back orders are included, the output parameters are lot size, shortage quantity, and related cost. From Eqs. (1.1), (1.2), and (1.3), the output parameters are:

$$Q^* = \sqrt{\frac{2DC(K+I)}{IK}}$$

$$J^* = \sqrt{\frac{2DCI}{K(K+I)}}$$

$$E^* = \sqrt{\frac{2DCIK}{(K+I)}}$$

It is of interest to analyze increase/reduction in the carrying cost.

4.4.1 *Carrying Cost Increase*

With reference to an increase in the inventory carrying cost, the following effects may be observed:

(i) Decrease in the lot size
(ii) Increase in the maximum stock out quantity
(iii) Increase in the total related cost

Example 4.3
Consider the following set of input and output parameters:

D	C	I	K	Q	E	J
600	30	10	200	61.4817	585.54	2.93

Implement the increase in carrying cost (I) as follows:

% Increase in I	5%	10%	15%	20%	25%	30%
I	10.5	11	11.5	12	12.5	13

The decreased lot size and the increased values of J and E are shown in Table 4.5.

In order to generalize, a decrease in the lot size with reference to an increased I_1:

Table 4.5 Effects on parameters with stock out corresponding to holding cost increase

% Increase in I	5%	10%	15%	20%	25%	30%
I	10.5	11	11.5	12	12.5	13
Q	60.07	58.76	57.54	56.39	55.32	54.31
% Decrease in Q	2.3%	4.4%	6.4%	8.3%	10.0%	11.7%
J	3	3.06	3.13	3.2	3.25	3.31
% Increase in J	2.4%	4.4%	6.8%	9.2%	10.9%	13.0%
E	599.29	612.66	625.7	638.4	650.79	662.9
% Increase in E	2.3%	4.6%	6.9%	9.0%	11.1%	13.2%

$$= \sqrt{\frac{2DC(K+I)}{IK}} - \sqrt{\frac{2DC(K+I_1)}{I_1K}}$$

$$= \sqrt{\frac{2DC(K+I)}{IK}}\left[1 - \sqrt{\frac{I(K+I_1)}{I_1(K+I)}}\right]$$

$$\text{Increase in the optimum stock out units} = \sqrt{\frac{2DCI_1}{K(K+I_1)}} - \sqrt{\frac{2DCI}{K(K+I)}}$$

$$= \sqrt{\frac{2DCI}{K(K+I)}}\left[\sqrt{\frac{I_1(K+I)}{I(K+I_1)}} - 1\right]$$

And, increase in the total related cost:

$$= \sqrt{\frac{2DCI_1K}{(K+I_1)}} - \sqrt{\frac{2DCIK}{(K+I)}}$$

$$= \sqrt{\frac{2DCIK}{(K+I)}}\left[\sqrt{\frac{I_1(K+I)}{I(K+I_1)}} - 1\right]$$

The derived results are also summarized in Table 4.6.

Table 4.6 Results with reference to an increased holding cost along with stock out

Decrease in the lot size	$\sqrt{\frac{2DC(K+I)}{IK}}\left[1 - \sqrt{\frac{I(K+I_1)}{I_1(K+I)}}\right]$
% Decrease in the lot size	$1 - \sqrt{\frac{I(K+I_1)}{I_1(K+I)}}$
Additional related cost	$\sqrt{\frac{2DCIK}{(K+I)}}\left[\sqrt{\frac{I_1(K+I)}{I(K+I_1)}} - 1\right]$
% Increase in cost	$\sqrt{\frac{I_1(K+I)}{I(K+I_1)}} - 1$
Increase in the stock out units	$\sqrt{\frac{2DCI}{K(K+I)}}\left[\sqrt{\frac{I_1(K+I)}{I(K+I_1)}} - 1\right]$
% Increase in the stock out units	$\sqrt{\frac{I_1(K+I)}{I(K+I_1)}} - 1$

4.4.2 Carrying Cost Reduction

Carrying cost might get decreased because of certain changes in the operations or overall economy.

Example 4.4
Following are the input and output data:

D	*C*	*I*	*K*	*Q*	*E*	*J*
600	30	10	200	61.4817	585.54	2.9277

Analyze the effects with the following reduction in I:

% Decrease in I	5%	10%	15%	20%	25%	30%
I	9.5	9	8.5	8	7.5	7

The effects are represented by Table 4.7. These are:

(a) Procurement lot size increases.
(b) Optimum stock out units decrease.
(c) Related cost decreases.

With this estimation, it will be useful to plan for additional storage capacity with the lower stock out level. Customer satisfaction level might improve with lower level of stock outs. Overall cost savings are also beneficial.

In order to generalize, an increase in the lot size with reference to a decreased I_1:

$$= \sqrt{\frac{2DC(K+I_1)}{I_1 K}} - \sqrt{\frac{2DC(K+I)}{IK}}$$

$$= \sqrt{\frac{2DC(K+I)}{IK}}\left[\sqrt{\frac{I(K+I_1)}{I_1(K+I)}} - 1\right]$$

Table 4.7 Effects on parameters with stock out corresponding to holding cost decrease

% Decrease in I	5%	10%	15%	20%	25%	30%
I	9.5	9	8.5	8	7.5	7
Q	63.00	64.65	66.45	68.41	70.57	72.96
% Increase in Q	2.48%	5.16%	8.08%	11.27%	14.78%	18.67%
J	2.86	2.78	2.71	2.63	2.55	2.47
% Decrease in J	2.42%	4.90%	7.47%	10.13%	12.88%	15.73%
E	571.39	556.82	541.78	526.23	510.14	493.44
% Decrease in E	2.42%	4.90%	7.47%	10.13%	12.88%	15.73%

Table 4.8 Results with reference to a reduced holding cost along with stock out

Increase in the lot size	$\sqrt{\frac{2DC(K+I)}{IK}}\left[\sqrt{\frac{I(K+I_1)}{I_1(K+I)}}-1\right]$
% Increase in the lot size	$\sqrt{\frac{I(K+I_1)}{I_1(K+I)}}-1$
Reduction in the related cost	$\sqrt{\frac{2DCIK}{(K+I)}}\left[1-\sqrt{\frac{I_1(K+I)}{I(K+I_1)}}\right]$
% Reduction in cost	$1-\sqrt{\frac{I_1(K+I)}{I(K+I_1)}}$
Decrease in the stock out units	$\sqrt{\frac{2DCI}{K(K+I)}}\left[1-\sqrt{\frac{I_1(K+I)}{I(K+I_1)}}\right]$
% Decrease in the stock out units	$1-\sqrt{\frac{I_1(K+I)}{I(K+I_1)}}$

$$\text{Decrease in the optimum stock out units} = \sqrt{\frac{2DCI}{K(K+I)}} - \sqrt{\frac{2DCI_1}{K(K+I_1)}}$$

$$= \sqrt{\frac{2DCI}{K(K+I)}}\left[1-\sqrt{\frac{I_1(K+I)}{I(K+I_1)}}\right]$$

And, reduction in the total related cost:

$$= \sqrt{\frac{2DCIK}{(K+I)}} - \sqrt{\frac{2DCI_1K}{(K+I_1)}}$$

$$= \sqrt{\frac{2DCIK}{(K+I)}}\left[1-\sqrt{\frac{I_1(K+I)}{I(K+I_1)}}\right]$$

Whenever a decrease in the inventory holding cost is observed, such results can be used for the situation. Every company has different input/operational data. These data can be substituted in the related formula, and the effects can be estimated for further necessary action/plan. The derived results are also summarized in Table 4.8.

4.5 Interaction of Holding Cost with Other Parameters

Because of an increase in the inventory holding cost, the following effects occur:

(a) Decrease in the lot size
(b) Increase in the cost

Decrease in the lot size is represented by Fig. 4.2 when demand rate is similar.

There are situations when variation in the lot size might not be preferred. This is because of the well-set transportation economy/material-handling convenience among other reasons as shown in Fig. 4.3. One possible option to keep similar lot size may be an if it is possible.

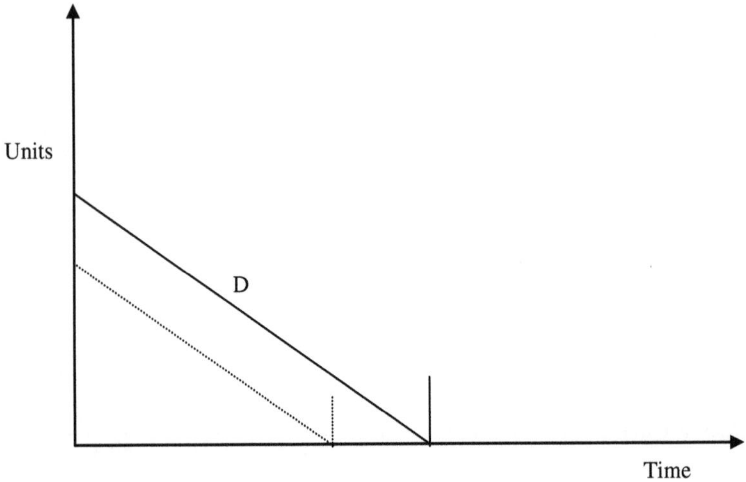

Fig. 4.2 Decrease in the lot size

Fig. 4.3 Possible reasons for keeping similar lot size

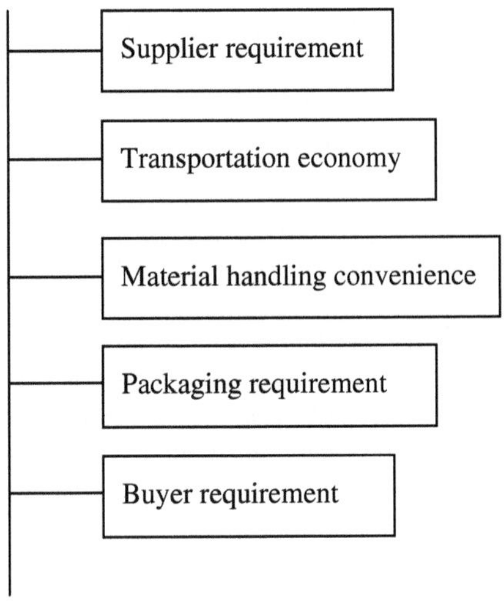

Example 4.5
An available set of information is as follows:

Annual demand, D = 600 units
Ordering cost, C = ₹30
Annual inventory carrying cost per unit, I = ₹10

The optimal output parameters are given by:

$$Q^* = \sqrt{\frac{2DC}{I}} = 60 \text{ units}$$
$$E^* = \sqrt{2DCI} = ₹600$$

Now, if inventory holding or carrying cost is increased by 10%, then:

$$I_1 = ₹11$$

Revised output parameters are:

$$Q = 57.21$$
$$E = ₹629.29$$

In order to have similar lot size:

$$\sqrt{\frac{2 \times 600 \times 30}{10}} = \sqrt{\frac{2 \times D_1 \times 30}{11}}$$

Or

$$D_1 = 660 \text{ units}$$

In some cases, an increase in demand is also possible by little effort such as the revival of customer relationship among other factors. This will also bring additional profit gain by way of selling larger quantities. Similar percentage demand increase would be necessary (i.e., equivalent to percentage holding cost increase) for obtaining similar lot size.

In case where objective is to keep similar cost levels, then the ordering cost may be lowered along with an increase in the holding cost, i.e.:

$$600 = \sqrt{2 \times 600 \times C_1 \times 11}$$

Or

$$C_1 = ₹27.27$$

When shortages are encountered (Fig. 4.4), the following effects may be observed with reference to an increase in the inventory carrying cost:

(i) Decrease in the lot size
(ii) Increase in the maximum stock out quantity
(iii) Increase in the total related cost

Example 4.6
Assume the following parameters:

Annual demand, $D = 600$ units
Ordering cost, $C = ₹30$

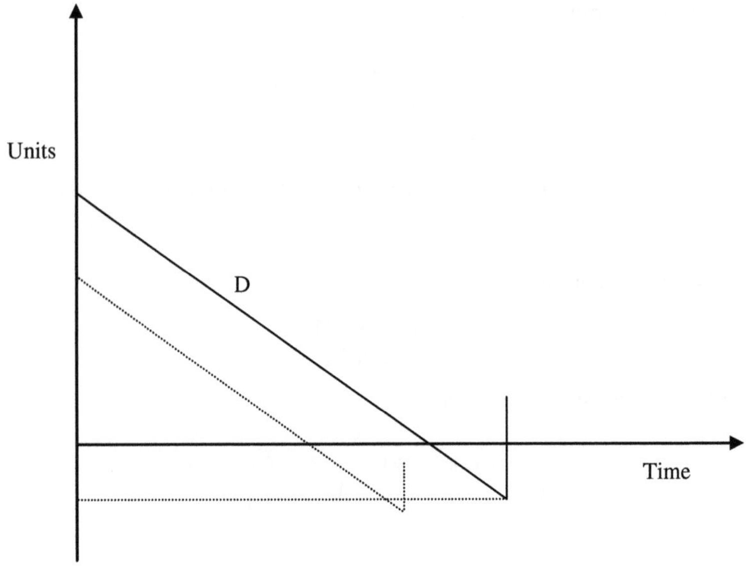

Fig. 4.4 Decrease in the lot size with stock outs

Annual inventory carrying cost per unit, I = ₹10
Annual stock out cost per unit, K = ₹200

Using the Eq. (1.11),

$$\begin{aligned} Q^* &= \sqrt{\frac{2DC(K+I)}{IK}} \\ &= \sqrt{\frac{2 \times 600 \times 30(200+10)}{10 \times 200}} \\ &= 61.48\,\text{units} \end{aligned}$$

From the Eq. (1.12), optimum stock out units:

$$\begin{aligned} J^* &= \sqrt{\frac{2DCI}{K(K+I)}} \\ &= \sqrt{\frac{2 \times 600 \times 30 \times 10}{200(200+10)}} \\ &= 2.93\,\text{units} \end{aligned}$$

And the total related annual cost from Eq. (1.13),

$$E^* = \sqrt{\frac{2DCIK}{(K+I)}}$$

$$= \sqrt{\frac{2 \times 600 \times 30 \times 10 \times 200}{(200+10)}}$$

$$= ₹585.54$$

Now:

$$I_1 = ₹11$$

The revised lot size is 58.76.
In order to have similar lot size, an increased demand can be found as:

$$\sqrt{\frac{2 \times D_1 \times 30 \times (200+11)}{11 \times 200}} = 61.48$$

Or

$$D_1 = 656.84$$

For a general approach, let:

P = % increase in holding cost
M = % increase in demand

Therefore:

$$I_1 = \left(1 + \frac{P}{100}\right)I$$

$$D_1 = \left(1 + \frac{M}{100}\right)D$$

For similar lot size:

$$\sqrt{\frac{2D_1C(K+I_1)}{I_1K}} = \sqrt{\frac{2DC(K+I)}{IK}}$$

Or

$$\frac{2D_1C(K+I_1)}{I_1K} = \frac{2DC(K+I)}{IK}$$

Or

$$\frac{2D(1+M/100)C[K+I(1+P/100)]}{I(1+P/100)K} = \frac{2DC(K+I)}{IK}$$

Or

$$\frac{(1+M/100)[K+I(1+P/100)]}{(1+P/100)} = (K+I)$$

Or

$$\left(1+\frac{M}{100}\right) = \frac{(K+I)(1+P/100)}{K+I(1+P/100)}$$

Or

$$\frac{M}{100} = \frac{(K+I)(1+P/100)}{K+I(1+P/100)} - 1$$

Or

$$\frac{M}{100} = \frac{K(1+P/100)+I(1+P/100) - K - I(1+P/100)}{K+I(1+P/100)}$$

Or

$$\frac{M}{100} = \frac{K(1+P/100) - K}{K+I(1+P/100)}$$

Or

$$\frac{M}{100} = \frac{(KP/100)}{K+I(1+P/100)}$$

Or

$$M = \frac{KP}{K+I(1+P/100)}$$

For the current situation, variation of M is shown in Table 4.9 with respect to P. The values are lower than the corresponding values of P.

Table 4.9 Variation of M with respect to P including stock out for similar lot size

S. No.	P	$M = \frac{KP}{K+I(1+P/100)}$
1	10	9.48
2	20	18.87
3	30	28.17
4	40	37.38
5	50	46.51

Example 4.7

Consider the data of previous example.

If the objective is to have similar cost level, then the reduced ordering cost can be found as:

$$\sqrt{\frac{2 \times 600 \times C_1 \times 11 \times 200}{(200+11)}} = 585.54$$

Or

$$C_1 = ₹27.40$$

In order to generalize, the additional notation is as follows:

P = % increase in holding cost
M = % reduction in the ordering cost

Therefore:

$$I_1 = \left(1 + \frac{P}{100}\right) I$$

$$C_1 = \left(1 - \frac{M}{100}\right) C$$

For similar total cost:

$$\sqrt{\frac{2DC_1 I_1 K}{(K+I_1)}} = \sqrt{\frac{2DCIK}{(K+I)}}$$

Or

$$\frac{C_1 I_1}{(K+I_1)} = \frac{CI}{(K+I)}$$

Or

$$C_1 = \frac{CI(K + I_1)}{I_1(K + I)}$$

Or

$$\left(1 - \frac{M}{100}\right) = \frac{I(K + I_1)}{I_1(K + I)}$$

Or

$$1 - \frac{M}{100} = \frac{K + I(1 + P/100)}{(1 + P/100)(K + I)}$$

Or

$$\frac{M}{100} = 1 - \frac{K + I(1 + P/100)}{(1 + P/100)(K + I)}$$

Or

$$\frac{M}{100} = \frac{(1 + P/100)(K + I) - K - I(1 + P/100)}{(1 + P/100)(K + I)}$$

Or

$$\frac{M}{100} = \frac{(1 + P/100)K - K}{(1 + P/100)(K + I)}$$

Or

$$\frac{M}{100} = \frac{(KP/100)}{(1 + P/100)(K + I)}$$

Or

$$M = \frac{KP}{(1 + P/100)(K + I)}$$

For the current situation, variation of M is shown in Table 4.10 with respect to P. The values are comparatively less sensitive corresponding to the higher values of P.

Table 4.10 Variation of M with respect to P including stock out for similar total cost

S. No.	P	$M = \frac{KP}{(1+P/100)(K+I)}$
1	10	8.66
2	20	15.87
3	30	21.98
4	40	27.21
5	50	31.75

Variation in M for similar cost level is lower in comparison with that related to a similar lot size. This is because:

$$\frac{KP}{(1+P/100)(K+I)} < \frac{KP}{K+I(1+P/100)}$$

Or

$$\frac{1}{(1+P/100)(K+I)} < \frac{1}{K+I(1+P/100)}$$

Or

$$(1+P/100)(K+I) > K+I(1+P/100)$$

Or

$$(1+P/100)K+(1+P/100)I > K+I(1+P/100)$$

Or

$$(1+P/100)K > K$$

Or

$$(1+P/100) > 1$$

And that is true.

Example 4.8
With the following parameters:

Annual demand, D = 600 units
Ordering cost, C = ₹30
Annual inventory carrying cost per unit, I = ₹10
Annual stock out cost per unit, K = ₹200

The total related annual cost from Eq. (1.13),

$$E^* = \sqrt{\frac{2DCIK}{(K+I)}}$$

$$= \sqrt{\frac{2\times 600\times 30\times 10\times 200}{(200+10)}}$$

$$= ₹585.54$$

Now, if ordering cost is increased by 20%, then the reduced inventory carrying cost in order to have similar total cost is obtained as:

$$\sqrt{\frac{2\times 600\times 36\times 200I_1}{(200+I_1)}} = 585.54$$

Or

$$8640000I_1 = 342857.09I_1 + 68571418$$

Or

$$8297142.9I_1 = 68571418$$

Or

$$I_1 = 8.26$$

In order to generalize, consider the following notation:

P = % increase in ordering cost
M = % reduction in the holding cost

Therefore:

$$C_1 = \left(1 + \frac{P}{100}\right)C$$

$$I_1 = \left(1 - \frac{M}{100}\right)I$$

For similar total cost:

$$\sqrt{\frac{2DC_1I_1K}{(K+I_1)}} = \sqrt{\frac{2DCIK}{(K+I)}}$$

Or

$$\frac{C_1 I_1}{(K + I_1)} = \frac{CI}{(K + I)}$$

Or

$$C_1 I_1 (K + I) = CI(K + I_1)$$

Or

$$C_1 I_1 K + C_1 I I_1 = CIK + CII_1$$

Or

$$I_1 [IC_1 + KC_1 - CI] = CIK$$

Or

$$I_1 = \frac{CIK}{C_1(I + K) - CI}$$

Or

$$\left(1 - \frac{M}{100}\right) = \frac{K}{(1 + P/100)(I + K) - I}$$

Or

$$\frac{M}{100} = 1 - \frac{K}{(1 + P/100)I + (1 + P/100)K - I}$$

Or

$$\frac{M}{100} = 1 - \frac{K}{(IP/100) + (1 + P/100)K}$$

Or

$$\frac{M}{100} = \frac{(IP/100) + (KP/100)}{(IP/100) + (1 + P/100)K}$$

Table 4.11 Variation of M (concerning I) with respect to P for similar total cost

S. No.	P	$M = \frac{(I+K)P}{(IP/100)+(1+P/100)K}$
1	10	9.50
2	20	17.36
3	30	23.95
4	40	29.58
5	50	34.43

Or

$$M = \frac{(I+K)P}{(IP/100)+(1+P/100)K}$$

For the current situation, variation of M (concerning I) is shown in Table 4.11 with respect to P. The values are lower than P and also comparatively less sensitive corresponding to the higher values of P.

While comparing with Table 4.10, the values of M are higher. This is because:

$$\frac{(I+K)P}{(IP/100)+(1+P/100)K} - \frac{KP}{(1+P/100)(K+I)} > 0$$

Or

$$\frac{(I+K)P[(1+P/100)(K+I)] - KP[(IP/100)+(1+P/100)K]}{[(IP/100)+(1+P/100)K][(1+P/100)(K+I)]} > 0$$

As the denominator of the above expression is positive, the present objective of positivity of L.H.S. can still be proved if numerator is positive, i.e.:

$$(I+K)P[(1+P/100)(K+I)] - KP[(IP/100)+(1+P/100)K] > 0$$

Or

$$(I+K)^2(1+P/100) - (KIP/100) - K^2(1+P/100) > 0$$

Or

$$(1+P/100)\left[(I+K)^2 - K^2\right] - (KIP/100) > 0$$

Or

$$(1+P/100)\left[I^2+2KI\right] - (KIP/100) > 0$$

Or

$$(1 + P/100)[I + 2K] - (KP/100) > 0$$

Or

$$I + 2K + (IP/100) + (2KP/100) - (KP/100) > 0$$

Or

$$I + 2K + (IP/100) + (KP/100) > 0$$

As L.H.S. is positive, this is proved.

Example 4.9
With the following parameters:

Annual demand, D = 600 units
Ordering cost, C = ₹30
Annual inventory carrying cost per unit, I = ₹10
Annual stock out cost per unit, K = ₹200

Optimum procurement lot size is obtained as:

$$\begin{aligned} Q &= \sqrt{\frac{2DC(K + I)}{IK}} \\ &= \sqrt{\frac{2 \times 600 \times 30(200 + 10)}{10 \times 200}} \\ &= 61.48 \text{ units} \end{aligned}$$

Now, if the change in parameter is triggered by a reduction in holding cost by 20%, then:

$$I_1 = ₹8$$

The revised lot size is 68.41.

In order to have similar lot size, a reduced ordering cost can be found as:

$$\sqrt{\frac{2 \times 600 \times C_1 \times (200 + 8)}{8 \times 200}} = 61.48$$

Or

$$C_1 = 24.23$$

For a general approach, let:

P = % reduction in holding cost
M = % reduction in ordering cost

Therefore:

$$I_1 = \left(1 - \frac{P}{100}\right) I$$
$$C_1 = \left(1 - \frac{M}{100}\right) C$$

For similar lot size:

$$\sqrt{\frac{2DC_1(K + I_1)}{I_1 K}} = \sqrt{\frac{2DC(K + I)}{IK}}$$

Or

$$C_1(K + I_1)I = C(K + I)I_1$$

Or

$$C_1 = \frac{C(K + I)I_1}{(K + I_1)I}$$

Or

$$1 - \frac{M}{100} = \frac{(K + I)(1 - P/100)}{K + I(1 - P/100)}$$

Or

$$\frac{M}{100} = \frac{K + I(1 - P/100) - (K + I)(1 - P/100)}{K + I(1 - P/100)}$$

Or

$$\frac{M}{100} = \frac{K - K(1 - P/100)}{K + I(1 - P/100)}$$

Or

$$\frac{M}{100} = \frac{(KP/100)}{K + I(1 - P/100)}$$

Or

$$M = \frac{KP}{K + I(1 - P/100)}$$

Table 4.12 Variation of *M* (regarding *C*) with *P* (regarding *I*) for similar lot size

S. No.	P	$M = \frac{KP}{K + I(1 - P/100)}$
1	10	9.57
2	20	19.23
3	30	28.98
4	40	38.83
5	50	48.78

For the current situation, variation of *M* is shown in Table 4.12 with respect to *P*. The values are comparatively lower than the corresponding values of *P*.

Concerning the interaction of parameters, a change is initiated by a parameter and the management response is with a variation in another parameter. However, a response can be in the form of a combination of suitable variation in the two parameters instead of one.

4.6 Response with a Combination of Parameters

In case where an aim is to have similar total cost and change is initiated by a demand increase, then options might include a reduction in either holding cost or an ordering cost. However, the management response can be associated with a combination of variation in the two parameters, i.e., ordering and holding costs.

4.6.1 Without Stock Outs

Implementation of the two-parameter variation as the response is illustrated with the next example.

Example 4.10
In the basic case, without stock out, consider the following information:

Annual demand, $D = 600$ units
Ordering cost, $C = 30$
Annual inventory carrying cost per unit, $I = 10$

The optimal output parameters are given by:

$$Q^* = \sqrt{\frac{2DC}{I}} = 60 \text{ units}$$

$$E^* = \sqrt{2DCI} = ₹600$$

Now, consider the demand increase by 10%, i.e., $D_1 = 660$ and the associated cost:

$$E = ₹629.29$$

With the increased demand also, if the objective is to maintain a previous total cost level, then the desired ordering cost, C_1, can be obtained as follows:

$$\sqrt{2 \times 660 \times C_1 \times 10} = 600$$

Or

$$2 \times 660 \times C_1 \times 10 = 600 \times 600$$

Or

$$C_1 = \frac{600 \times 600}{2 \times 660 \times 10}$$

Or

$$C_1 = ₹27.27$$

Percentage reduction in the ordering cost is approximately 9.09%, i.e., less than the demand increase by 10%.

In practice, there might be a limitation on furthermore reduction in the ordering cost. For instance, assume that the ordering cost can be reduced by maximum 5% only. Now, it becomes necessary to explore a reduction in another parameter, i.e., the holding cost. Reduced value of holding cost is obtained as follows:

$$\sqrt{2 \times 660 \times 28.5 \times I_1} = 600$$

Or

$$2 \times 660 \times 28.5 \times I_1 = 600 \times 600$$

Or

$$I_1 = \frac{600 \times 600}{2 \times 660 \times 28.5}$$

Or

$$I_1 = 9.57$$

Now, the management response can be with a combination of the following parameters:

(i) Ordering cost
(ii) Holding cost

When a change is triggered by an increase in demand by 10%, then the response might be as follows in order to retain the similar total cost:

C_1 = ₹28.5, i.e., 5% reduction
I_1 = ₹9.57, i.e., 4.3% decrease

In order to generalize, the notation is as follows:

P = % increase in demand
M = % reduction in the ordering cost
N = % reduction in the holding cost

Now:

$$\sqrt{2D_1C_1I_1} = \sqrt{2DCI}$$

Or

$$2D_1C_1I_1 = 2DCI$$

Or

$$D_1C_1I_1 = DCI$$

Or

$$\left(1+\frac{P}{100}\right)D\left(1-\frac{M}{100}\right)C\left(1-\frac{N}{100}\right)I = DCI$$

Or

$$\left(1+\frac{P}{100}\right)\left(1-\frac{M}{100}\right)\left(1-\frac{N}{100}\right) = 1$$

For a specific value of P, variety of combinations of M and N can be generated. Suitable combination of values for M and N may be implemented depending on the feasibility.

For example, with P = 10:

$$\left(1-\frac{M}{100}\right)\left(1-\frac{N}{100}\right) = \frac{1}{1.1}$$

Or

$$1-\frac{N}{100} = \frac{(1/1.1)}{1-(M/100)}$$

Table 4.13 Values of N (concerning I) corresponding to M (concerning C) for P = 10

S. No.	M	$N = \frac{(10/1.1)-M}{(1-M/100)}$
1	1	8.17
2	2	7.24
3	3	6.28
4	4	5.30
5	5	4.31
6	6	3.29
7	7	2.25
8	8	1.19

Or

$$\frac{N}{100} = 1 - \frac{(1/1.1)}{(1 - M/100)}$$

Or

$$\frac{N}{100} = \frac{1 - (M/100) - (1/1.1)}{(1 - M/100)}$$

Or

$$\frac{N}{100} = \frac{(0.1/1.1) - (M/100)}{(1 - M/100)}$$

Or

$$N = \frac{(10/1.1) - M}{(1 - M/100)}$$

The values of N (for P = 10) are shown in Table 4.13 corresponding to the different values of M.

Example 4.11
With reference to the data of previous example, lot size is given by:

$$Q^* = \sqrt{\frac{2DC}{I}} = 60\,\text{units}$$

If a change is initiated by a reduction in ordering cost say by 10% and an aim is to have similar lot size, then the management response can be a combination of:

(i) Potential increase in demand
(ii) Reduction in carrying cost

For a generalization, consider:

P = % reduction in ordering cost
M = % increase in the demand
N = % reduction in the holding cost

Now:

$$\sqrt{\frac{2D_1C_1}{I_1}} = \sqrt{\frac{2DC}{I}}$$

Or

$$DCI_1 = D_1C_1I$$

Or

$$DCI\left(1 - \frac{N}{100}\right) = DCI\left(1 + \frac{M}{100}\right)\left(1 - \frac{P}{100}\right)$$

Or

$$\left(1 - \frac{N}{100}\right) = \left(1 + \frac{M}{100}\right)\left(1 - \frac{P}{100}\right)$$

Or

$$\frac{N}{100} = 1 - \left(1 + \frac{M}{100}\right)\left(1 - \frac{P}{100}\right)$$

Or

$$\frac{N}{100} = 1 - \left(1 - \frac{P}{100} + \frac{M}{100} - \frac{MP}{10000}\right)$$

Or

$$\frac{N}{100} = \frac{P}{100} + \frac{MP}{10000} - \frac{M}{100}$$

Or

$$N = P - M + \frac{MP}{100}$$

Table 4.14 Values of N (concerning I) corresponding to M (concerning D) for $P = 10$

S. No.	M	$N = 10 - 0.9M$
1	1	9.1
2	2	8.2
3	3	7.3
4	4	6.4
5	5	5.5
6	6	4.6
7	7	3.7
8	8	2.8

Or

$$N = P - M(1 - P/100)$$

The values of N (for $P = 10$) are shown in Table 4.14 corresponding to the different values of M.

4.6.2 *With Stock Outs*

Along with the stock outs, the analysis is made either for similar total cost or for similar lot size.

Example 4.12
With the following parameters:

Annual demand, $D = 600$ units
Ordering cost, $C =$ ₹30
Annual inventory carrying cost per unit, $I =$ ₹10
Annual stock out cost per unit, $K =$ ₹200

The total related annual cost:

$$E^* = \sqrt{\frac{2DCIK}{(K+I)}}$$

$$= \sqrt{\frac{2 \times 600 \times 30 \times 10 \times 200}{(200+10)}}$$

$$= ₹585.54$$

Now, if the demand has increased by 20%, then the reduced ordering cost for a similar total cost is obtained as:

$$\sqrt{\frac{2 \times 720 \times C_1 \times 200 \times 10}{(200 + 10)}} = 585.54$$

Or

$$C_1 = ₹25$$

That is a reduction of approx. 16.67%.

However, if it is not possible to reduce the ordering cost by more than 8% (say), then a combination of ordering and carrying cost might be considered. For this case (considering max. reduction in C as 8%):

$$\sqrt{\frac{2 \times 720 \times 27.6 \times 200 \times I_1}{(200 + I_1)}} = 585.54$$

Or

$$I_1 = 9.02$$

That is, 9.8% reduction in holding cost.

For a general approach:

P = % increase in demand
M = % reduction in the ordering cost
N = % reduction in the holding cost

Now:

$$\sqrt{\frac{2D_1C_1I_1K}{(K + I_1)}} = \sqrt{\frac{2DCIK}{(K + I)}}$$

Or

$$\frac{D_1C_1I_1}{(K + I_1)} = \frac{DCI}{(K + I)}$$

Or

$$D_1C_1I_1(K + I) = DCI(K + I_1)$$

Or

$$(1 + P/100)(1 - M/100)(1 - N/100)(K + I) = K + I(1 - N/100)$$

Or

$$(1+P/100)(1-M/100)(1-N/100)(K+I)-I(1-N/100)=K$$

Or

$$(1-N/100)[(1+P/100)(1-M/100)(K+I)-I]=K$$

Or

$$1-\frac{N}{100}=\frac{K}{(1+P/100)(1-M/100)(K+I)-I}$$

Or

$$\frac{N}{100}=\frac{(1+P/100)(1-M/100)(K+I)-(I+K)}{(1+P/100)(1-M/100)(K+I)-I}$$

Or

$$N=\frac{100(K+I)[(1+P/100)(1-M/100)-1]}{(1+P/100)(1-M/100)(K+I)-I}$$

With the use of example data and for $P = 20$:

$$N=\frac{21000[0.2-0.012M]}{252(1-M/100)-10}$$

Or

$$N=\frac{4200-252M}{242-2.52M}$$

The values of N (for $P = 20$) are shown in Table 4.15 corresponding to the different values of M.

Table 4.15 N (for I) with stock out corresponding to M (for C) concerning $P = 20$

S. No.	M	$N=\frac{4200-252M}{242-2.52M}$
1	1	16.49
2	2	15.60
3	3	14.69
4	4	13.76
5	5	12.82
6	6	11.85
7	7	10.86
8	8	9.84

Example 4.13
With the following parameters:

Annual demand, D = 600 units
Ordering cost, C = ₹30
Annual inventory carrying cost per unit, I = ₹10
Annual stock out cost per unit, K = ₹200

Optimum procurement lot size is obtained as:

$$Q = \sqrt{\frac{2DC(K+I)}{IK}}$$
$$= \sqrt{\frac{2 \times 600 \times 30(200+10)}{10 \times 200}}$$
$$= 61.48 \text{ units}$$

Now, if the change in parameter is triggered by a reduction in ordering cost by 20%, then:

$$C_1 = ₹24$$

The revised lot size is 54.99.

In order to have similar lot size, a potentially increased demand can be found as:

$$\sqrt{\frac{2 \times D_1 \times 24 \times (200+10)}{10 \times 200}} = 61.48$$

Or

$$D_1 = 749.96$$

That is 24.99% increase in demand.

However, if it is not possible to increase the demand by more than 10% (say), then a combination of demand and carrying cost might be considered. For this case (considering max. increase in D as 10%):

$$\sqrt{\frac{2 \times 660 \times 24 \times (200+I_1)}{200 I_1}} = 61.48$$

Or

$$I_1 = 8.75$$

That is 12.5% reduction in the carrying cost.

Percentage variation in the parameters can be linked with the efforts needed to implement. Larger percentage increase in demand requires larger level of effort. Beyond certain extent, it might not be possible also. In the present example, 24.99%

increase in demand might be replaced by a combination of variation of the parameters such as:

(i) 10% increase in demand
(ii) 12.5% reduction in the carrying cost

This can be implemented if it is feasible. A general approach is useful in order to generate several combinations in terms of different level of variation in the parameters. An appropriate level of variation might be selected by the company depending on ease of implementation. An ease in implementation may differ from company to company.

For a generalized approach, let:

P = % reduction in the ordering cost
M = % increase in demand
N = % reduction in the holding cost

Now:

$$\sqrt{\frac{2D_1C_1(K+I_1)}{I_1K}} = \sqrt{\frac{2DC(K+I)}{IK}}$$

Or

$$DC(K+I)I_1 = D_1C_1(K+I_1)I$$

Or

$$DCI(1-N/100)(K+I) = DCI(1-P/100)(1+M/100)[K+I(1-N/100)]$$

Or

$$\begin{aligned}&(1-N/100)(K+I)-(1-P/100)(1+M/100)I(1-N/100)\\&\quad=(1-P/100)(1+M/100)K\end{aligned}$$

Or

$$(1-N/100)[(K+I)-I(1-P/100)(1+M/100)] = (1-P/100)(1+M/100)K$$

Or

$$1-\frac{N}{100} = \frac{(1-P/100)(1+M/100)K}{(K+I)-I(1-P/100)(1+M/100)}$$

Or

$$\frac{N}{100} = \frac{(K+I) - I(1 - P/100)(1+M/100) - (1 - P/100)(1+M/100)K}{(K+I) - I(1 - P/100)(1+M/100)}$$

Or

$$\frac{N}{100} = \frac{K+I[1 - (1 - P/100)(1+M/100)] - (1 - P/100)(1+M/100)K}{K+I[1 - (1 - P/100)(1+M/100)]}$$

Or

$$\frac{N}{100} = \frac{K+I - (K+I)(1 - P/100)(1+M/100)}{K+I[1 - (1 - P/100)(1+M/100)]}$$

Or

$$\frac{N}{100} = \frac{(K+I)[1 - (1 - P/100)(1+M/100)]}{K+I[1 - (1 - P/100)(1+M/100)]}$$

Or

$$N = \frac{100(K+I)[1 - (1 - P/100)(1+M/100)]}{K+I[1 - (1 - P/100)(1+M/100)]}$$

With the use of example data and for $P = 20$:

$$N = \frac{21000[1 - 0.8(1+M/100)]}{200+10[1 - 0.8(1+M/100)]}$$

Or

$$N = \frac{21000[0.2 - (0.8M/100)]}{200+10[0.2 - (0.8M/100)]}$$

Or

$$N = \frac{4200 - 168M}{202 - 0.08M}$$

The values of N (for $P = 20$) are shown in Table 4.16 corresponding to the different values of M.

Table 4.16 N (for I) with stock out corresponding to M (for D) concerning $P = 20$

S. No.	M	$N = \frac{4200-168M}{202-0.08M}$
1	1	19.97
2	2	19.14
3	3	18.32
4	4	17.49
5	5	16.67
6	6	15.84
7	7	15.01
8	8	14.18
9	9	13.35
10	10	12.52

4.7 Carrying Cost Estimation for Various Products

Variety of products may belong to different types of industry such as:

(a) FMCG
(b) Pharmaceutical
(c) Food/agricultural products
(d) Hi-tech

4.7.1 *FMCG*

A fast-moving consumer goods company might supply the products to several locations. While estimating the inventory carrying costs, different factors (that vary from location to location) should be analyzed such as:

(i) Availability of storage space
(ii) Own or hired storage space
(iii) Power charges
(iv) Spoilage

Because of the environment at a particular location, some of products may spoil, thus contributing toward higher inventory carrying cost. If the storage space is not easily available at a particular location or it is costly, inventory carrying cost goes up. The contribution of power charges may be on account of maintaining air-conditioned or suitable temperature environment. Lower power charges may help in lowering the inventory holding costs.

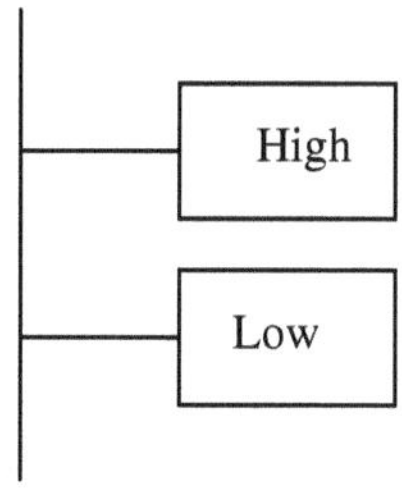

Fig. 4.5 Stock level

4.7.2 *Pharmaceutical*

Stock level for the medicines is shown in Fig. 4.5. Inventory carrying cost also depends on the level of medicine stock maintained at various locations in the chain, i.e., right from the producing companies to the retailers. Although stock level has a link with the level of expected demand/consumption pattern, this correspondence might not be observed in certain cases that vary from location to location. Additional aspects might be as follows which are also interrelated:

(a) Expiry date
(b) Remaining shelf life

Expiry date may be determined and declared by the producing companies. However, remaining shelf life at a particular location may depend on:

(i) Storage at intermediate places
(ii) Transportation between any two points at several segments
(iii) Storage time
(iv) Transportation time
(v) Transportation delay

If stock level is high, the expiry date might be over before consumption by the patient or before reaching the end consumer. This may be treated as unnecessary spoilage contributing toward a higher carrying cost. On the other hand, if stock level is lower, there is a possibility of suffering of the consumers for want of medicine. A very appropriate stock level should be determined which justifies a suitable inventory holding costs at each location in the medicine chain.

4.7.3 *Food/Agricultural Products*

Food grains are stored at strategic locations so that shortages may not occur. A decision for keeping the inventory at such locations is in the view of costs and transit time to take these to the place of consumption. In case of food products or the output of food-processing companies, an appropriate environment is necessary to maintain in order to extend the shelf life of items also including the vegetables and fruits. Remaining shelf life should be sufficient enough to encourage the potential customers to buy it. These aspects provide additional help in estimating the inventory holding costs at various levels in the food chain or agricultural chain. The following aspects might also be considered:

(i) Space requirement:
If enough space is not available to keep the food grains and other such products, there might be a chance of unnecessary spoilage.
(ii) Appropriate environment during storage:
Depending on the need, refrigeration and air-conditioned environment might be necessary in order to preserve the food/agricultural products.
(iii) Appropriate environment during transit:
When transit time or transportation time is longer, temperature maintenance during the transit also might be necessary for preserving the products and improving the shelf life. However, in case of certain output of the food-processing companies, temperature maintenance is necessary throughout, i.e., during storage as well as transit, even if transit or transportation time might be comparatively short.
(iii) Waste disposal:
Timely and effective waste disposal helps in inventory assessment and its valuation along with synchronization of appropriate inventory stock and precise requirement of different consumers at different locations.

4.7.4 Hi-Tech

As the hi-tech products may include computer and telecommunication equipment, product life cycle is shorter and therefore higher level of inventory might not be preferred depending on the case. However, fluctuation in the inventory holding costs might be due to the price decrease of some of the components. There is price reduction in certain component because of cheaper option available as a result of R&D/innovation efforts. This may also result in an overall price decrease of the equipment leading to an inventory carrying cost reduction. Therefore, a periodic estimation of holding cost considering such factors among others helps in effective decision making that pertains to production and inventory control.

Chapter 5
Procurement Cost

While procuring certain item, the following components may comprise the associated cost:

(i) Purchase cost
(ii) Transportation cost
(iii) Relevant loading/unloading

For further analysis, there is a need to estimate the procurement cost per unit. Depending on the situation, the procurement cost may represent either purchase cost or the sum of purchase and transportation cost among other associated factors. For the purpose of analyzing the procurement cost, the formulation requires certain adjustment. The annual inventory carrying cost per unit (I) is the multiplication of unit procurement cost and an annual inventory carrying cost fraction.

Let:

P = procurement cost per unit
F = annual inventory carrying cost fraction

Refer Eq. (1.3), i.e., the total annual cost:

$$E = \left[\frac{D}{Q}\right].C + \left[\frac{Q}{2}\right].I$$

For the present situation, annual procurement cost, i.e., PD may also be added. Additionally,

Annual inventory carrying cost, $I = PF$.

Because in the present scenario, procurement cost is to be specifically analyzed, the total annual cost:

S. Sharma, *Inventory Parameters*, DOI 10.1007/978-981-10-4545-5_5

$$E = PD + \left[\frac{D}{Q}\right].C + \left[\frac{Q}{2}\right].PF \tag{5.1}$$

However, the first component (i.e., PD) does not contain Q, and therefore, it does not affect the solution particularly lot size. Following the procedure discussed in Chap. 1:

$$Q^* = \sqrt{\frac{2DC}{PF}} \tag{5.2}$$

$$E^* = PD + \sqrt{2DCPF} \tag{5.3}$$

When stock out situation is incorporated, the following results may be obtained [With reference to Eqs. (1.11), (1.12), and (1.13)]:

$$Q^* = \sqrt{\frac{2DC(K+PF)}{PFK}} \tag{5.4}$$

$$J^* = \sqrt{\frac{2DCPF}{K(K+PF)}} \tag{5.5}$$

$$E^* = PD + \sqrt{\frac{2DCPFK}{(K+PF)}} \tag{5.6}$$

5.1 Procurement Cost Reduction

Overall procurement cost might get reduced because of one or more of the following reasons:

(a) A price discount is offered by the supplier anticipating an increase in the demand of their products.
(b) There is a reduction in the taxes imposed on the prices.
(c) Transportation cost gets decreased as a result of fuel price reduction.
(d) With the help of modern loading/unloading practices, the associated costs get reduced.
(e) An overall supply of certain item is increased in the market, and therefore, price of that item is reduced.

Example 5.1
Assume the following parameters:

Annual demand, D = 600 units

Ordering cost, C = ₹30
Procurement cost per unit, P = ₹120
Annual inventory carrying cost fraction, F = 0.1

Using Eq. (5.2),

$$
\begin{aligned}
Q^* &= \sqrt{\frac{2DC}{PF}} \\
&= \sqrt{\frac{2 \times 600 \times 30}{120 \times 0.1}} \\
&= 54.77\,\text{units}
\end{aligned}
$$

And the total related annual cost from Eq. (5.3),

$$
\begin{aligned}
E^* &= PD + \sqrt{2DCPF} \\
&= (120 \times 600) + \sqrt{2 \times 600 \times 30 \times 120 \times 0.1} \\
&= ₹\,72657.27
\end{aligned}
$$

In case of a procurement cost reduction, the lot size and total cost get affected. If such potential effects are known well in advance, it helps the management in furthermore planning.

Example 5.2
With reference to the previous example, the set for input and output data is as follows:

F	D	C	P	Q	E
0.1	600	30	120	54.77	72657.27

Implement a reduction in procurement cost as follows:

% reduction in P	5%	10%	15%	20%	25%	30%
P	114	108	102	96	90	84

Effects on parameters are shown in Table 5.1. Procurement lot size increases whereas total related cost decreases. In such situation, arrangement for additional space would be necessary. However, cost savings may be utilized elsewhere if needed. Variation in total cost is more than that in the procurement lot size.

Let:

A = % decrease in unit procurement cost:

$$P_1 = \left(1 - \frac{A}{100}\right)P$$

Table 5.1 Effects on parameters with respect to procurement cost reduction

% reduction in P	5%	10%	15%	20%	25%	30%
P	114	108	102	96	90	84
Q	56.20	57.74	59.41	61.24	63.25	65.47
% increase in Q	2.60%	5.41%	8.47%	11.80%	15.47%	19.52%
E	69040.62	65423.54	61805.97	58187.88	54569.21	50949.91
% reduction in E	4.98%	9.96%	14.93%	19.91%	24.90%	29.88%

$$\begin{aligned}\text{Increase in the lot size} &= \sqrt{\frac{2DC}{FP_1}} - \sqrt{\frac{2DC}{FP}} \\ &= \sqrt{\frac{2DC}{F\left(1-\frac{A}{100}\right)P}} - \sqrt{\frac{2DC}{FP}} \\ &= \sqrt{\frac{2DC}{FP}}\left[\sqrt{\frac{1}{\left(1-\frac{A}{100}\right)}} - 1\right]\end{aligned}$$

And,

$$\%\,\text{increase in } Q = \sqrt{\frac{1}{\left(1-\frac{A}{100}\right)}} - 1$$

Now,

$$\begin{aligned}\text{Annual cost reduction} &= PD + \sqrt{2DCFP} - DP_1 - \sqrt{2DCFP_1} \\ &= D(P-P_1) + \sqrt{2DCF}\left[\sqrt{P} - \sqrt{P_1}\right] \\ &= DP\left[1-\left(1-\frac{A}{100}\right)\right] + \sqrt{2DCF}\left[\sqrt{P} - \sqrt{\left(1-\frac{A}{100}\right)P}\right] \\ &= \frac{DPA}{100} + \sqrt{2DCFP}\left[1-\sqrt{\left(1-\frac{A}{100}\right)}\right]\end{aligned}$$

And:

$$\%\,\text{decrease in } E = \frac{(DPA/100) + \sqrt{2DCFP}\{1-\sqrt{1-(A/100)}\}}{DP + \sqrt{2DCFP}}$$

These generalized results are also summarized in Table 5.2.

Table 5.2 Results with reference to % decrease in unit procurement cost

Increase in the lot size	$\sqrt{\frac{2DC}{FP}}\left[\sqrt{\frac{1}{\left(1-\frac{A}{100}\right)}}-1\right]$
% increase in the lot size	$\sqrt{\frac{1}{\left(1-\frac{A}{100}\right)}}-1$
Annual cost reduction	$\frac{DPA}{100}+\sqrt{2DCFP}\left[1-\sqrt{\left(1-\frac{A}{100}\right)}\right]$
% decrease in annual cost	$\frac{(DPA/100)+\sqrt{2DCFP}\left\{1-\sqrt{1-(A/100)}\right\}}{DP+\sqrt{2DCFP}}$

5.2 Procurement Cost Increase

Procurement cost may get increased because of one or more of the following reasons:

(a) An overall higher demand of certain in the market may result into higher price of that item.
(b) Higher wages associated with human resources involved in the related loading/unloading tasks increase the procurement cost.
(c) Increased level of taxes imposed on the price of certain item.
(d) A fuel price increase contributing toward an increased transportation cost.

Example 5.3
Consider the base data of Example 5.1. Analyze if an increase in P is as follows:

% Increase in P	5%	10%	15%	20%	25%	30%
P	126	132	138	144	150	156

Procurement cost per unit, P, is increased by the given percentage, and approximate values are obtained such as:

(i) Q
(ii) % decrease in Q
(iii) E
(iv) % increase in E

Table 5.3 represents the variation in parameters with reference to an increase in P. The management will need to plan for additional expenditure because of total cost increase. However, some space would become free due to a decrease in procurement lot size. Variation in cost is more than that in the procurement lot size.

Let:

A = % increase in unit procurement cost

$$P_1 = \left(1 + \frac{A}{100}\right)P$$

Table 5.3 Effects on parameters with respect to procurement cost increase

% increase in P	5%	10%	15%	20%	25%	30%
P	126	132	138	144	150	156
Q	53.45	52.22	51.08	50.00	48.99	48.04
% decrease in Q	2.41%	4.65%	6.75%	8.71%	10.56%	12.29%
E	76273.50	79889.35	83504.84	87120.00	90734.85	94349.40
% increase in E	4.98%	9.95%	14.93%	19.91%	24.88%	29.86%

$$\begin{aligned}\text{Decrease in the lot size} &= \sqrt{\frac{2DC}{FP}} - \sqrt{\frac{2DC}{FP_1}}\\ &= \sqrt{\frac{2DC}{FP}} - \sqrt{\frac{2DC}{F\left(1+\frac{A}{100}\right)P}}\\ &= \sqrt{\frac{2DC}{FP}}\left[1 - \sqrt{\frac{1}{\left(1+\frac{A}{100}\right)}}\right]\end{aligned}$$

And:

$$\%\,\text{decrease in } Q = 1 - \sqrt{\frac{1}{\left(1+\frac{A}{100}\right)}}$$

Now:

$$\begin{aligned}\text{Additional annual cost incurred} &= D(P_1 - P) + \sqrt{2DCFP_1} - \sqrt{2DCFP}\\ &= \left(\frac{DPA}{100}\right) + \sqrt{2DCF}\left[\sqrt{P_1} - \sqrt{P}\right]\\ &= \left(\frac{DPA}{100}\right) + \sqrt{2DCF}\left[\sqrt{\left(1+\frac{A}{100}\right)P} - \sqrt{P}\right]\\ &= \left(\frac{DPA}{100}\right) + \sqrt{2DCFP}\left[\sqrt{\left(1+\frac{A}{100}\right)} - 1\right]\end{aligned}$$

And:

$$\%\,\text{increase in } E = \frac{(DPA/100) + \sqrt{2DCFP}\left\{\sqrt{1+(A/100)} - 1\right\}}{DP + \sqrt{2DCFP}}$$

The obtained generalized results are summarized in Table 5.4.

Table 5.4 Results with reference to % increase in unit procurement cost

Decrease in the lot size	$\sqrt{\frac{2DC}{FP}}\left[1-\sqrt{\frac{1}{\left(1+\frac{A}{100}\right)}}\right]$
% decrease in the lot size	$1-\sqrt{\frac{1}{\left(1+\frac{A}{100}\right)}}$
Additional related cost	$\left(\frac{DPA}{100}\right)+\sqrt{2DCFP}\left[\sqrt{\left(1+\frac{A}{100}\right)}-1\right]$
% increase in cost	$\frac{(DPA/100)+\sqrt{2DCFP}\left\{\sqrt{1+(A/100)}-1\right\}}{DP+\sqrt{2DCFP}}$

Considering the numerical results, % variation in total cost seems to be slightly lower than that in the previous case, i.e., procurement cost decrease. This is because:

$$\frac{(DPA/100)+\sqrt{2DCFP}\{\sqrt{1+(A/100)}-1\}}{DP+\sqrt{2DCFP}} < \frac{(DPA/100)+\sqrt{2DCFP}\{1-\sqrt{1-(A/100)}\}}{DP+\sqrt{2DCFP}}$$

Or

$$\sqrt{\left(1+\frac{A}{100}\right)}-1<1-\sqrt{\left(1-\frac{A}{100}\right)}$$

Or

$$\sqrt{\left(1+\frac{A}{100}\right)}+\sqrt{\left(1-\frac{A}{100}\right)}<2$$

Or

$$\left[\sqrt{\left(1+\frac{A}{100}\right)}+\sqrt{\left(1-\frac{A}{100}\right)}\right]^2<4$$

Or

$$1+\frac{A}{100}+1-\frac{A}{100}+2\sqrt{\left(1+\frac{A}{100}\right)\left(1-\frac{A}{100}\right)}<4$$

Or

$$2+2\sqrt{1-\left(\frac{A}{100}\right)^2}<4$$

Or

$$2\sqrt{1-\left(\frac{A}{100}\right)^2}<2$$

Or

$$\sqrt{1-\left(\frac{A}{100}\right)^2}<1$$

And that is true for all practical values of $A > 0$ and $A < 100$.

% variation in lot size in previous case is higher than the present, i.e., the procurement cost increases. In other way,

$$\sqrt{\frac{1}{(1-\frac{A}{100})}}-1>1-\sqrt{\frac{1}{(1+\frac{A}{100})}}$$

Or

$$2<\sqrt{\frac{1}{(1-A/100)}}+\sqrt{\frac{1}{(1+A/100)}}$$

Or

$$2\left[\sqrt{1-(A/100)^2}\right]<\sqrt{(1+A/100)}+\sqrt{(1-A/100)}$$

Or

$$4\left[1-(A/100)^2\right]<1+(A/100)+1-(A/100)+2\sqrt{1-(A/100)^2}$$

Or

$$2\left[1-(A/100)^2\right]<1+\sqrt{1-(A/100)^2}$$

Or

$$2-2(A/100)^2<1+\sqrt{1-(A/100)^2}$$

Or

$$1 - (A/100)^2 < (A/100)^2 + \sqrt{1 - (A/100)^2}$$

The above expression is valid even if:

$$1 - (A/100)^2 < \sqrt{1 - (A/100)^2}$$

It is true for all practical values of A. This is because L.H.S. is between 0 and 1 (i.e., greater than zero, and less than one), and square root of any such value is greater than that value.

5.3 Stock Out Inclusion

In order to know the effects of variation in the procurement cost under stock out scenario, shortages have been included.

Example 5.4
Consider the following parameters:

Annual demand, D = 600 units
Ordering cost, C = ₹30
Procurement cost per unit, P = ₹120
Annual inventory carrying cost fraction, F = 0.1
Annual stock out cost per unit, K = ₹200

Using Eq. (5.4),

$$\begin{aligned} Q^* &= \sqrt{\frac{2DC(K + PF)}{PFK}} \\ &= \sqrt{\frac{2 \times 600 \times 30[200 + (120 \times 0.1)]}{120 \times 0.1 \times 200}} \\ &= 56.39 \text{ units} \end{aligned}$$

From the Eq. (5.5), optimum stock out units:

$$\begin{aligned} J^* &= \sqrt{\frac{2DCPF}{K(K + PF)}} \\ &= \sqrt{\frac{2 \times 600 \times 30 \times 120 \times 0.1}{200[200 + (120 \times 0.1)]}} \\ &= 3.19 \text{ units} \end{aligned}$$

And the total related annual cost from Eq. (5.6),

$$E^* = PD + \sqrt{\frac{2DCPFK}{(K+PF)}}$$

$$= (120\times600) + \sqrt{\frac{2\times600\times30\times120\times0.1\times200}{[200+(120\times0.1)]}}$$

$$= ₹72638.39$$

With reference to the previous situation, ordering quantity increased but overall cost has decreased.

5.3.1 *Decreased Purchase Cost*

In case of the decreased purchase/procurement cost from P to P_1, illustration is made with the help of following example:

Example 5.5
Consider the base data of Example 5.4 as given below:

F	*D*	*C*	*P*	*K*	*Q*	*E*	*J*
0.1	600	30	120	200	56.39149	72638.39	3.191971

Analyze if a decrease in P is as follows:

% decrease in *P*	5%	10%	15%	20%	25%	30%
P	114	108	102	96	90	84

Implementation of the purchase/procurement cost reduction is represented by Table 5.5 which shows variation in the output parameters.

An increase in the procurement batch size because of a reduced P_1:

$$= \sqrt{\frac{2DC(K+FP_1)}{KFP_1}} - \sqrt{\frac{2DC(K+FP)}{KFP}}$$

$$= \sqrt{\frac{2DC(K+FP)}{KFP}}\left[\sqrt{\frac{P(K+FP_1)}{P_1(K+FP)}} - 1\right]$$

Table 5.5 Effects on parameters with stock out corresponding to purchase cost decrease

% decrease in P	5%	10%	15%	20%	25%	30%
P	114	108	102	96	90	84
Q	57.77	59.27	60.90	62.69	64.65	66.83
% increase in Q	2.45%	5.11%	8.00%	11.17%	14.65%	18.50%
J	3.12	3.04	2.96	2.87	2.78	2.69
% decrease in J	2.39%	4.86%	7.41%	10.05%	12.78%	15.61%
E	69023.11	65407.36	61791.09	58174.26	54556.82	50938.71
% decrease in E	4.98%	9.95%	14.93%	19.91%	24.89%	29.87%

$$\text{Reduction in the optimum stock out units} = \sqrt{\frac{2DCFP}{K(K+FP)}} - \sqrt{\frac{2DCFP_1}{K(K+FP_1)}}$$
$$= \sqrt{\frac{2DCFP}{K(K+FP)}}\left[1 - \sqrt{\frac{P_1(K+FP)}{P(K+FP_1)}}\right]$$

And, decrease in the total related cost:

$$= DP + \sqrt{\frac{2DCKFP}{(K+FP)}} - DP_1 - \sqrt{\frac{2DCKFP_1}{(K+FP_1)}}$$
$$= D(P-P_1) + \sqrt{\frac{2DCKFP}{(K+FP)}}\left[1 - \sqrt{\frac{P_1(K+FP)}{P(K+FP_1)}}\right]$$

The derived results are also summarized in Table 5.6.

Table 5.6 Results with reference to a reduced purchase cost along with stock out

Increase in the lot size	$\sqrt{\frac{2DC(K+FP)}{KFP}}\left[\sqrt{\frac{P(K+FP_1)}{P_1(K+FP)}} - 1\right]$
% increase in the lot size	$\sqrt{\frac{P(K+FP_1)}{P_1(K+FP)}} - 1$
Decrease in the related cost	$D(P-P_1) + \sqrt{\frac{2DCKFP}{(K+FP)}}\left[1 - \sqrt{\frac{P_1(K+FP)}{P(K+FP_1)}}\right]$
Reduction in the stock out units	$\sqrt{\frac{2DCFP}{K(K+FP)}}\left[1 - \sqrt{\frac{P_1(K+FP)}{P(K+FP_1)}}\right]$
% reduction in the stock out units	$1 - \sqrt{\frac{P_1(K+FP)}{P(K+FP_1)}}$

5.3.2 *Increased Purchase Cost*

With an increased purchase/procurement cost, the effects on the parameters are illustrated using the example as follows:

Example 5.6
Consider the base data of Example 5.4 as given below:

F	*D*	*C*	*P*	*K*	*Q*	*E*	*J*
0.1	600	30	120	200	56.39149	72638.39	3.191971

Analyze if an increase in *P* is as follows:

% increase in *P*	5%	10%	15%	20%	25%	30%
P	126	132	138	144	150	156

After implementing the procurement cost increase, the variation in output parameters is shown in Table 5.7. These are:

(i) Reduction in lot size
(ii) Increase in optimum stock out level
(iii) Increase in total relevant cost

With the reduction in lot size, some space would be available for certain other use if needed. However, there is additional expenditure that needs to be arranged by the company under consideration.

In order to generalize, a decrease in the lot size with reference to an increased P_1:

Table 5.7 Effects on parameters with stock out corresponding to purchase cost increase

% increase in *P*	5%	10%	15%	20%	25%	30%
P	126	132	138	144	150	156
Q	55.11	53.92	52.81	51.77	50.79	49.88
% decrease in *Q*	2.27%	4.38%	6.35%	8.20%	9.93%	11.55%
J	3.27	3.34	3.41	3.48	3.54	3.61
% increase in *J*	2.32%	4.59%	6.79%	8.93%	11.02%	13.06%
E	76253.24	79867.67	83481.71	87095.40	90708.75	94321.78
% increase in *E*	4.98%	9.95%	14.93%	19.90%	24.88%	29.85%

$$= \sqrt{\frac{2DC(K+FP)}{KFP}} - \sqrt{\frac{2DC(K+FP_1)}{KFP_1}}$$

$$= \sqrt{\frac{2DC(K+FP)}{KFP}}\left[1 - \sqrt{\frac{P(K+FP_1)}{P_1(K+FP)}}\right]$$

% decrease in the lot size = $\left[1 - \sqrt{\frac{P(K+FP_1)}{P_1(K+FP)}}\right]$

Increase in the optimum stock out units $= \sqrt{\frac{2DCFP_1}{K(K+FP_1)}} - \sqrt{\frac{2DCFP}{K(K+FP)}}$

$$= \sqrt{\frac{2DCFP}{K(K+FP)}}\left[\sqrt{\frac{P_1(K+FP)}{P(K+FP_1)}} - 1\right]$$

% increase in the stock out units = $\left[\sqrt{\frac{P_1(K+FP)}{P(K+FP_1)}} - 1\right]$

And, increase in the total related cost:

$$= DP_1 + \sqrt{\frac{2DCKFP_1}{(K+FP_1)}} - DP - \sqrt{\frac{2DCKFP}{(K+FP)}}$$

$$= D(P_1 - P) + \sqrt{\frac{2DCKFP}{(K+FP)}}\left[\sqrt{\frac{P_1(K+FP)}{P(K+FP_1)}} - 1\right]$$

The derived results are also summarized in Table 5.8.

Table 5.8 Results with reference to an increased purchase cost along with stock out

Decrease in the lot size	$\sqrt{\frac{2DC(K+FP)}{KFP}}\left[1 - \sqrt{\frac{P(K+FP_1)}{P_1(K+FP)}}\right]$
% decrease in the lot size	$1 - \sqrt{\frac{P(K+FP_1)}{P_1(K+FP)}}$
Additional related cost	$D(P_1 - P) + \sqrt{\frac{2DCKFP}{(K+FP)}}\left[\sqrt{\frac{P_1(K+FP)}{P(K+FP_1)}} - 1\right]$
Increase in the stock out units	$\sqrt{\frac{2DCFP}{K(K+FP)}}\left[\sqrt{\frac{P_1(K+FP)}{P(K+FP_1)}} - 1\right]$
% increase in the stock out units	$\sqrt{\frac{P_1(K+FP)}{P(K+FP_1)}} - 1$

5.4 Interaction of Procurement Cost with Other Parameters

With procurement cost increase, the following effects take place:

(a) Reduction in lot size
(b) Increase in total cost

Example 5.7
With the use of the following parameters:

Annual demand, D = 600 units
Ordering cost, C = ₹30
Procurement cost per unit, P = ₹120
Annual inventory carrying cost fraction, F = 0.1

The output parameters are obtained as:

$$Q^* = \sqrt{\frac{2DC}{PF}}$$
$$= \sqrt{\frac{2 \times 600 \times 30}{120 \times 0.1}}$$
$$= 54.77 \text{ units}$$

And the total related annual cost:

$$E^* = PD + \sqrt{2DCPF}$$
$$= (120 \times 600) + \sqrt{2 \times 600 \times 30 \times 120 \times 0.1}$$
$$= ₹72657.27$$

With 10% increase in procurement cost:

P_1 = ₹132
Reduced lot size = 52.22
Increased total cost = ₹79889.35

(i) If an objective is to have similar lot size, then the reduced value of F is obtained as follows:

$$\sqrt{\frac{2 \times 600 \times 30}{132 \times F_1}} = 54.77$$

Or

$$F_1 = 0.091$$

For a general approach, let:
A = % increase in unit procurement cost

$$P_1 = \left(1 + \frac{A}{100}\right)P$$

M = % reduction in the value of F

$$F_1 = \left(1 - \frac{M}{100}\right)F$$

Now,

$$\sqrt{\frac{2DC}{P_1F_1}} = \sqrt{\frac{2DC}{PF}}$$

Or

$$P_1F_1 = PF \tag{5.7}$$

Or

$$\left(1 + \frac{A}{100}\right)P\left(1 - \frac{M}{100}\right)F = PF$$

Or

$$\left(1 + \frac{A}{100}\right)\left(1 - \frac{M}{100}\right) = 1$$

Or

$$\left(1 - \frac{M}{100}\right) = \frac{1}{(1 + A/100)}$$

Or

$$\frac{M}{100} = 1 - \frac{1}{(1 + A/100)}$$

Or

$$\frac{M}{100} = \frac{(A/100)}{(1 + A/100)}$$

Or

$$M = \frac{A}{(1 + A/100)} \tag{5.8}$$

(ii) If the focus is on total cost, then an option might be considered as follows:

Despite the increased procurement cost and subsequent total cost, it is always of interest to reduce the total cost to certain feasible level. A possible decrease in the annual inventory carrying cost fraction might be one option among others.

For instance, target total annual cost = ₹79800.

Now, the desired F_1 can be obtained as follows:

$$(132 \times 600) + \sqrt{2 \times 600 \times 30 \times 132F_1} = 79800$$

Or

$$\sqrt{2 \times 600 \times 30 \times 132F_1} = 600$$

Or

$$2 \times 600 \times 30 \times 132F_1 = 600 \times 600$$

Or

$$7920F_1 = 600$$

Or

$$F_1 = 0.076$$

Example 5.8
Consider the information from the previous Example:

P_1 = ₹132
Reduced lot size = 52.22
Increased total cost = ₹79889.35

In order to achieve the target total annual cost = ₹79800, another option might be a reduced ordering cost (C_1) that can be obtained as follows:

$$(132 \times 600) + \sqrt{2 \times 600 \times C_1 \times 132 \times 0.1} = 79800$$

Or

$$\sqrt{2 \times 600 \times C_1 \times 132 \times 0.1} = 600$$

Or

$$2 \times 600 \times C_1 \times 132 \times 0.1 = 600 \times 600$$

Or

$$26.4C_1 = 600$$

Or

$$C_1 = ₹22.73$$

In order to get a comparatively lower level of total cost in spite of an increase in procurement cost, the following options might be available:

(a) Reduction in the annual inventory carrying cost fraction
(b) Reduction in the ordering cost

Depending on the ease in implementation, any suitable reduction in parameter is possible. This may relate to storage or ordering process modification among other aspects. The discussed approach is also represented by Fig. 5.1. However, if it is possible to reduce both parameters simultaneously under the existing business/operational constraints, then the target cost itself might get revised. Such iterative nature of procedure is helpful in making continuous efforts to reduce the total cost in the event of a procurement cost increase concerning a product.

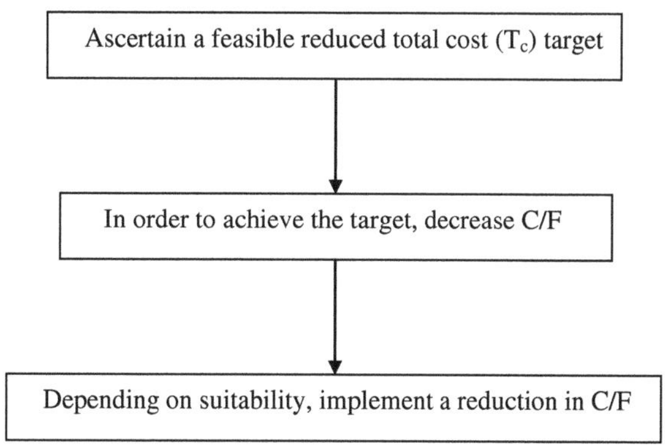

Fig. 5.1 Approach for reducing total cost

For instance, consider the case when it is possible to implement the parameters as follows:

$$C_1 = ₹22.73$$
$$F_1 = 0.076$$

Furthermore revision in the reduced total cost (T_c) target can be obtained as:

$$T_c = (132 \times 600) + \sqrt{2 \times 600 \times 22.73 \times 132 \times 0.076}$$

Or

$$T_c = ₹79{,}723$$

Example 5.9
With the base data as follows:

F	D	C	P	K	Q	E	J
0.1	600	30	120	200	56.39	72638.39	3.19

Implement $P_1 = ₹132$
Now:
The increased cost

$$E_1 = P_1D + \sqrt{\frac{2DCP_1FK}{(K + P_1F)}}$$

Or

$$E_1 = ₹79867.67$$

For instance, if the target annual cost = ₹79800, then the desired F_1 can be obtained as follows:

$$(132 \times 600) + \sqrt{\frac{2 \times 600 \times 30 \times 132 \times 200F_1}{200 + 132F_1}} = 79800$$

Or

$$\sqrt{\frac{2 \times 600 \times 30 \times 132 \times 200F_1}{200 + 132F_1}} = 600$$

Or

$$2 \times 600 \times 30 \times 132 \times 200F_1 = 600 \times 600(200 + 132F_1)$$

Or

$$9.0288 \times 10^8 F_1 = 72 \times 10^6$$

Or

$$F_1 = \frac{72}{902.88}$$

Or

$$F_1 = ₹0.0797$$

In order to achieve the target annual cost, another option might be to get the reduced ordering cost as follows:

$$(132 \times 600) + \sqrt{\frac{2 \times 600 \times C_1 \times 132 \times 200 \times 0.1}{200 + (132 \times 0.1)}} = 79800$$

Or

$$\frac{2 \times 600 \times C_1 \times 132 \times 200 \times 0.1}{200 + (132 \times 0.1)} = 600 \times 600$$

Or

$$C_1 = ₹24.23$$

In case where an aim is to have similar lot size, then the reduced F is obtained as follows:

$$\sqrt{\frac{2 \times 600 \times 30(200 + 132F_1)}{200 \times 132F_1}} = 56.39$$

Or

$$F_1 = 0.091$$

This value is similar to the Example 5.7 when stock outs were not allowed. This is because:

$$\sqrt{\frac{2DC(K + P_1F_1)}{P_1F_1K}} = \sqrt{\frac{2DC(K + PF)}{PFK}}$$

Or

$$\frac{K + P_1F_1}{P_1F_1} = \frac{K + PF}{PF}$$

Or

$$P_1F_1(K + PF) = PF(K + P_1F_1)$$

Or

$$P_1F_1K + P_1F_1PF = PFK + PFP_1F_1$$

Or

$$P_1F_1K = PFK$$

Or

$$P_1F_1 = PF$$

This is similar to Eq. (5.7), and therefore, Eq. (5.8) is applicable for the present stock out case also, i.e.,

$$M = \frac{A}{(1 + A/100)}$$

Chapter 6
Cycle Time

In the present context, cycle time may refer to the duration in which the procured quantity is consumed completely. After that, the procured quantity arrives again and similar cycle repeats.

6.1 Significance of Cycle Time

As shown in Fig. 6.1, quantity Q is procured, it is consumed in time T, and the cycle repeats itself. Frequency of ordering in a year increases if the cycle time T is reduced. Sometimes because of the buyer requirements, cycle time needs to be increased.

In the context of optimization, cycle time (C.T.) needs to be lowered (Fig. 6.2) in the following situations:

(i) In order to cater to an increased demand rate, cycle time needs to be lower.
(ii) Higher inventory carrying cost.
(iii) Lower ordering cost.

Additionally if the shelf life is getting over for an item, the cycle time for that item needs to be adjusted, i.e., needs to be lowered. This is because the item should be consumed within the cycle time.

A higher C.T. (Fig. 6.3) may be required for the following reasons:

(i) Decreased demand rate;
(ii) Lower inventory carrying cost;
(iii) Higher ordering cost

S. Sharma, *Inventory Parameters*, DOI 10.1007/978-981-10-4545-5_6

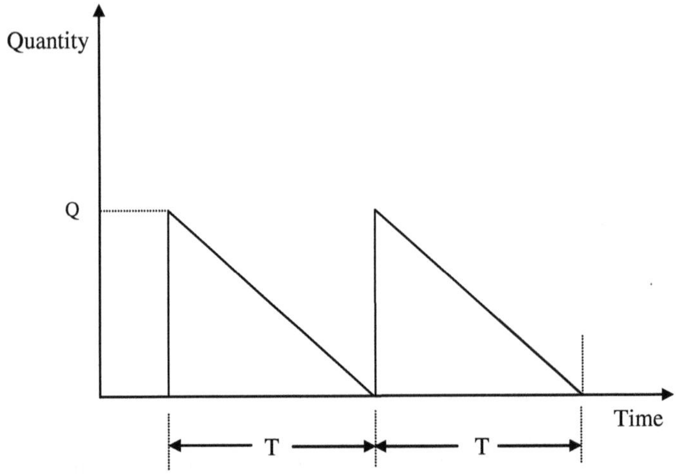

Fig. 6.1 Cycle time

Fig. 6.2 Possible reasons for lower C.T

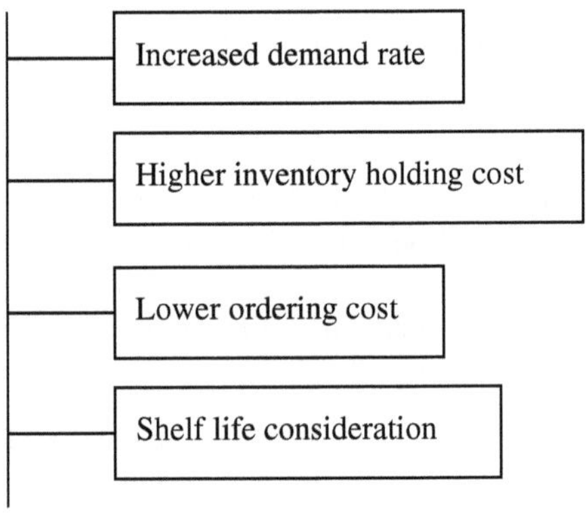

Fig. 6.3 Possible reasons for higher C.T

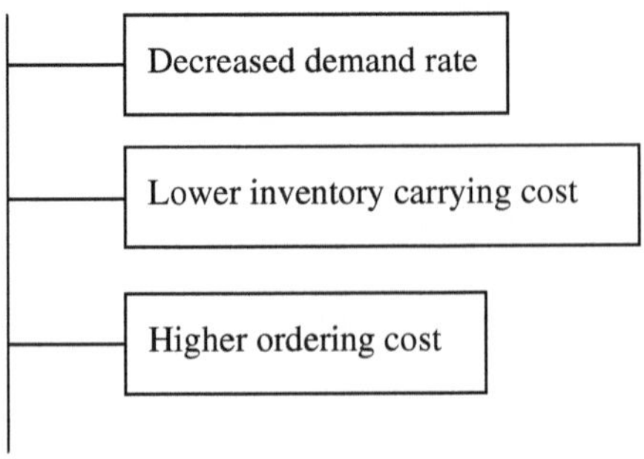

Because of the significance of cycle time, it becomes necessary to formulate in terms of T. Suppose that there are four procurement cycles in a year, then the cycle time:

$$T = \frac{1}{4} = 0.25\ \text{year.}$$

In other words, the number of orders in one year $= \frac{1}{T}$.

If the fixed ordering cost per unit order is C, then the annual ordering cost (AOC) is as follows:

$$\text{AOC} = \frac{C}{T} \tag{6.1}$$

In case where annual demand is 12000 items and the cycle time is 0.25 year, the procurement quantity per order will be:

$$12000 \times 0.25 = 3000$$

That is, the ordering quantity = DT.

In the beginning of the cycle time, available inventory is (DT) units, and it is zero at the end of the cycle time. Such inventory cycle continues throughout the year, therefore the average inventory is:

$$\frac{DT}{2}$$

If the annual inventory carrying cost is I per unit item, then the annual inventory carrying cost (AIC) is as follows:

$$\text{AIC} = \frac{DTI}{2} \tag{6.2}$$

Now the total annual cost (E) can be obtained by adding Eqs. (6.1) and (6.2), i.e.,

$$E = \text{AOC} + \text{AIC}$$

Or

$$E = \frac{C}{T} + \frac{DTI}{2} \tag{6.3}$$

Differentiating with respect to T and making it equivalent to zero:

$$\frac{DI}{2} - \frac{C}{T^2} = 0$$

Or

$$\frac{DI}{2} = \frac{C}{T^2}$$

Or

$$T^* = \sqrt{\frac{2C}{DI}} \tag{6.4}$$

In order to get the optimal annual cost in terms of C.T., value of DI from Eq. (6.4) may be substituted in Eq. (6.3).

$$E^* = \frac{C}{T} + \frac{T}{2} \cdot \frac{2C}{T^2}$$

Or

$$E^* = \frac{C}{T} + \frac{C}{T}$$

Or

$$E^* = \frac{2C}{T} \tag{6.5}$$

Example 6.1
Consider the following parameters:

Annual demand, D = 600 items
Ordering cost, C = ₹200
Annual inventory carrying cost per unit, I = ₹11

From Eq. (6.4), cycle time:

$$T = \sqrt{\frac{2 \times 200}{600 \times 11}}$$

Or

$$T = 0.246 \text{ year}$$

From Eq. (6.5), the total cost:

$$E = \frac{2 \times 200}{0.246}$$

Or

E = ₹ 1624.81

6.2 Cycle Time Variation

As discussed before, there might be a need to vary the cycle time. It may be increased or decreased. For example, it should be decreased if the shelf life of the concerned product is less than the cycle time.

Example 6.2
Assume the data of Example 6.1. In order to implement, cycle time is approximately 0.25 year. The number of procurement cycles in one year is:

$$\frac{1}{0.25} = 4$$

As the annual demand is of 600 products, 150 products would be procured in each cycle. Four cycles in one year are shown in Fig. 6.4.

Cycle time needs to be decreased in case where shelf life (or remaining shelf life after arrival of the procurement lot) of this product is 0.2 year (or slightly higher than this), i.e., less than the existing cycle time. Now, the reduced cycle time is 0.2 year as shown in Fig. 6.5.

In order to handle the shelf life of this product, the number of procurement cycles in one year is:

$$\frac{1}{0.2} = 5$$

As the annual demand is of 600 products, 120 products would be procured in each cycle. However, the total cost (with similar input parameters) will increase as follows:

$$E = \frac{2 \times 200}{0.2}$$

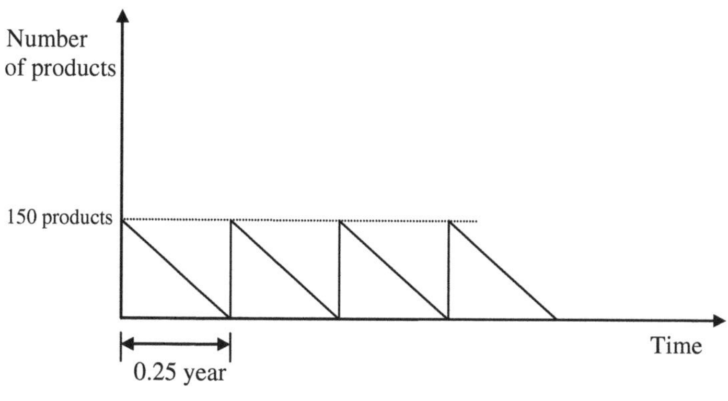

Fig. 6.4 Implementation of procurement cycle time in a year

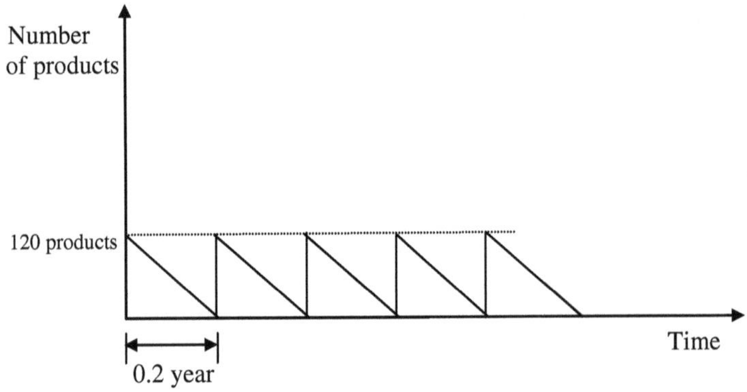

Fig. 6.5 Reduced procurement cycle time

Or

$$E = ₹\,2000$$

In general, Fig. 6.6 represents the reduction in cycle time (where only one cycle is shown) without affecting the input parameters, namely

(i) Demand rate
(ii) Ordering cost
(iii) Inventory carrying cost

Because of the input parameter change, cycle time varies and it needs to be optimized.

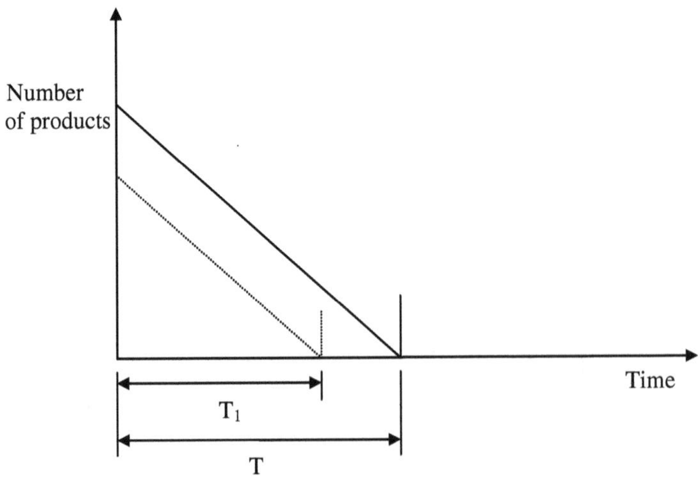

Fig. 6.6 Reduced cycle time (with similar input parameters)

6.2.1 *Reduced Cycle Time*

Optimal cycle time reduces because of:

(a) Higher demand
(b) Higher inventory holding cost
(c) Lower ordering cost

Let:
P = % variation in parameter

(a) **Higher demand**:

For the present discussion, P refers to the % increase in demand, therefore a higher demand:

$$D_1 = \left(1 + \frac{P}{100}\right)D$$

$$\begin{aligned}\text{Reduction in the cycle time} &= T - T_1 \\ &= \sqrt{\frac{2C}{DI}} - \sqrt{\frac{2C}{D_1 I}} \\ &= \sqrt{\frac{2C}{DI}} - \sqrt{\frac{2C}{\left(1 + \frac{P}{100}\right)DI}} \\ &= \sqrt{\frac{2C}{DI}}\left[1 - \sqrt{\frac{1}{\left(1 + \frac{P}{100}\right)}}\right]\end{aligned}$$

And:

$$\%\text{ reduction in C.T.} = 1 - \sqrt{\frac{1}{\left(1 + \frac{P}{100}\right)}}$$

This is also illustrated as follows:
Consider the data of Example 6.1, i.e.,

D	C	I	T	E
600	200	11	0.246	1624.81

After the implementation of an increase in demand, the approximate results are as follows:

% increase in D	5%	10%	15%	20%	25%	30%
D	630	660	690	720	750	780
T	0.24	0.235	0.23	0.225	0.22	0.216
% decrease in T	2.41%	4.65%	6.75%	8.71%	10.56%	12.29%
E	1664.93	1704.11	1742.41	1779.89	1816.59	1852.57
% increase in E	2.47%	4.88%	7.24%	9.54%	11.80%	14.02%

(b) **Higher inventory holding cost**:

Similar results can be obtained considering higher inventory holding cost:

$$I_1 = \left(1 + \frac{P}{100}\right)I$$

(c) **Lower ordering cost**:

The decreased cost of ordering:

$$C_1 = \left(1 - \frac{P}{100}\right)C$$

$$\begin{aligned}
\text{Reduction in the cycle time} &= T - T_1 \\
&= \sqrt{\frac{2C}{DI}} - \sqrt{\frac{2C_1}{DI}} \\
&= \sqrt{\frac{2C}{DI}} - \sqrt{\frac{2\left(1 - \frac{P}{100}\right)C}{DI}} \\
&= \sqrt{\frac{2C}{DI}}\left[1 - \sqrt{\left(1 - \frac{P}{100}\right)}\right]
\end{aligned}$$

And:

$$\%\ \text{reduction in C.T.} = 1 - \sqrt{\left(1 - \frac{P}{100}\right)}$$

For an illustration, consider the similar reference set of data, i.e.,:

D	C	I	T	E
600	200	11	0.246	1624.81

Table 6.1 Results for reduced cycle time

Variation in input parameter	Variation in optimal C.T.
Higher demand/inventory holding cost	Reduction in the cycle time = $\sqrt{\frac{2C}{DI}}\left[1-\sqrt{\frac{1}{\left(1+\frac{P}{100}\right)}}\right]$ % reduction in C.T. = $1-\sqrt{\frac{1}{\left(1+\frac{P}{100}\right)}}$
Lower ordering cost	Reduction in the cycle time = $\sqrt{\frac{2C}{DI}}\left[1-\sqrt{\left(1-\frac{P}{100}\right)}\right]$ % reduction in C.T. = $1-\sqrt{\left(1-\frac{P}{100}\right)}$

After using the lower ordering cost, the computational results are as follows:

% decrease in C	5%	10%	15%	20%	25%	30%
C	190	180	170	160	150	140
T	0.240	0.234	0.227	0.220	0.213	0.206
% decrease in T	2.53%	5.13%	7.80%	10.56%	13.40%	16.33%
E	1583.67	1541.43	1498.00	1453.27	1407.12	1359.41
% decrease in E	2.53%	5.13%	7.80%	10.56%	13.40%	16.33%

The obtained generalized results are summarized in Table 6.1.

6.2.2 *Increased Cycle Time*

There might be a need to increase the cycle time. For example, a wholesaler may tell the retailer to increase the cycle time as it suits the operational situation of the wholesaler. However, in general, input parameters at the retailer end alone might change, and accordingly optimal cycle time needs to be determined.

Optimal cycle time increases because of:

(a) Lower inventory holding cost
(b) Lower demand
(c) Higher ordering cost

Following similar procedure, increase in the cycle time and percentage increase in it can be derived with reference to input parameter variation.

Let:
T_1 = Increased cycle time.

(a) **Lower inventory holding cost**:

$$I_1 = \left(1 - \frac{P}{100}\right)I$$

$$\text{Increase in the C.T.} = T_1 - T$$
$$= \sqrt{\frac{2C}{DI_1}} - \sqrt{\frac{2C}{DI}}$$
$$= \sqrt{\frac{2C}{D(1-\frac{P}{100})I}} - \sqrt{\frac{2C}{DI}}$$
$$= \sqrt{\frac{2C}{DI}}\left[\sqrt{\frac{1}{(1-\frac{P}{100})}} - 1\right]$$

And:

$$\%\text{ increase in C.T.} = \sqrt{\frac{1}{(1-\frac{P}{100})}} - 1$$

From the previous discussion, the data set is as given below:

D	C	I	T	E
600	200	11	0.246	1624.81

The output parameters are provided along with lower inventory holding cost as follows:

% decrease in I	5%	10%	15%	20%	25%	30%
I	10.45	9.9	9.35	8.8	8.25	7.7
T	0.253	0.259	0.267	0.275	0.284	0.294
% increase in T	2.60%	5.41%	8.47%	11.80%	15.47%	19.52%
E	1583.67	1541.43	1498.00	1453.27	1407.12	1359.41
% decrease in E	2.53%	5.13%	7.80%	10.56%	13.40%	16.33%

(b) **Lower demand**:
Similar results can be obtained considering a decreased demand:

$$D_1 = \left(1 - \frac{P}{100}\right)D$$

(c) **Higher ordering cost**:

The increased cost of ordering:

$$C_1 = \left(1 + \frac{P}{100}\right)C$$

$$\begin{aligned}
\text{Increase in the cycle time} &= T_1 - T \\
&= \sqrt{\frac{2C_1}{DI}} - \sqrt{\frac{2C}{DI}} \\
&= \sqrt{\frac{2\left(1 + \frac{P}{100}\right)C}{DI}} - \sqrt{\frac{2C}{DI}} \\
&= \sqrt{\frac{2C}{DI}}\left[\sqrt{\left(1 + \frac{P}{100}\right)} - 1\right]
\end{aligned}$$

And:

$$\%\ \text{increase in C.T.} = \sqrt{\left(1 + \frac{P}{100}\right)} - 1$$

With the data set as given below:

D	C	I	T	E
600	200	11	0.246	1624.81

The results are found as follows:

% increase in C	5%	10%	15%	20%	25%	30%
C	210	220	230	240	250	260
T	0.252	0.258	0.264	0.270	0.275	0.281
% increase in T	2.47%	4.88%	7.24%	9.54%	11.80%	14.02%
E	1664.93	1704.11	1742.41	1779.89	1816.59	1852.57
% increase in E	2.47%	4.88%	7.24%	9.54%	11.80%	14.02%

The summarized results in general form are also represented by Table 6.2.

Table 6.2 Results for an increased cycle time

Variation in input parameter	Variation in optimal C.T.
Lower inventory holding cost/demand	Increase in the cycle time $= \sqrt{\frac{2C}{DI}}\left[\sqrt{\frac{1}{\left(1-\frac{P}{100}\right)}} - 1\right]$ % increase in C.T. $= \sqrt{\frac{1}{\left(1-\frac{P}{100}\right)}} - 1$
Higher ordering cost	Increase in the cycle time $= \sqrt{\frac{2C}{DI}}\left[\sqrt{\left(1 + \frac{P}{100}\right)} - 1\right]$ % increase in C.T. $= \sqrt{\left(1 + \frac{P}{100}\right)} - 1$

6.3 Backlogging Situation

As shown in Fig. 6.7, let J be the shortage quantity which is backlogged.

The stock out situation is for the time:

$$\frac{J}{D}$$

The average stock out is as follows:

$$\frac{J}{2}$$

Since there are $\frac{1}{T}$ cycles in a year, the annual shortage cost (ASC) is given by:

$$\frac{J}{2} \cdot \frac{J}{D} \cdot K \cdot \frac{1}{T}$$

Or:

$$\mathrm{ASC} = \frac{J^2 K}{2DT} \tag{6.6}$$

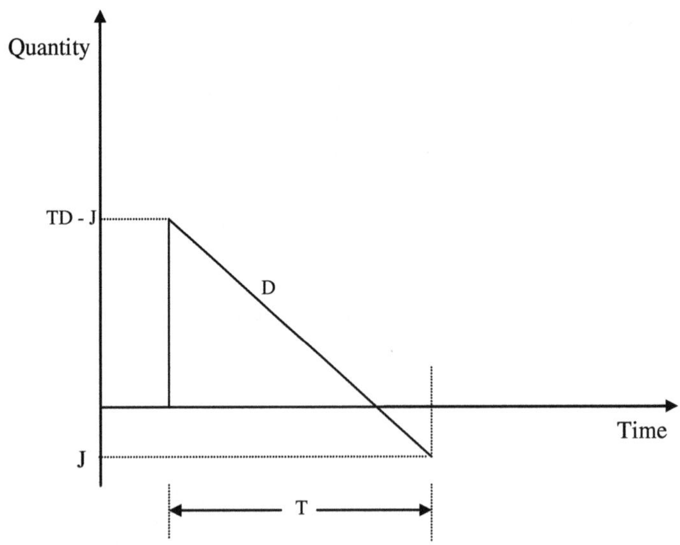

Fig. 6.7 Cycle time with backlogging

As all the shortages are assumed to be backlogged and the procurement ordering quantity is TD, the positive inventory exists for the time:

$$\frac{(TD - J)}{D}$$

Average positive inventory is $\frac{(TD-J)}{2}$, and there are $\frac{1}{T}$ cycles. The annual inventory carrying cost is expressed as:

$$\text{AIC} = \frac{(TD - J)}{2} \cdot \frac{(TD - J)}{D} \cdot I \cdot \frac{1}{T}$$

Or:

$$\text{AIC} = \frac{(TD - J)^2 I}{2TD} \tag{6.7}$$

Annual ordering cost is given by Eq. (6.1). Adding Eqs. (6.6), (6.7), and (6.1), total relevant cost is:

$$E = \frac{J^2 K}{2TD} + \frac{(TD - J)^2 I}{2TD} + \frac{C}{T}$$

Or

$$E = \frac{J^2 K}{2TD} + \frac{(T^2 D^2 - 2TDJ + J^2) I}{2TD} + \frac{C}{T}$$

Or

$$E = \frac{J^2 K}{2TD} + \frac{TDI}{2} - JI + \frac{J^2 I}{2TD} + \frac{C}{T}$$

Or

$$E = \frac{TDI}{2} + \frac{C}{T} + \frac{J^2 (K + I)}{2TD} - JI \tag{6.8}$$

Differentiating Eq. (6.8) partially with respect to J and equating to zero:

$$\frac{J(K + I)}{TD} - I = 0$$

Or

$$J = \frac{ITD}{(K+I)} \tag{6.9}$$

Putting the value of J in Eq. (6.8):

$$E = \frac{TDI}{2} + \frac{C}{T} + \frac{I^2T^2D^2}{(K+I)^2} \cdot \frac{(K+I)}{2TD} - \frac{I^2TD}{(K+I)}$$

Or

$$E = \frac{TDI}{2} + \frac{C}{T} + \frac{I^2TD}{2(K+I)} - \frac{I^2TD}{(K+I)}$$

Or

$$E = \frac{TDI}{2} + \frac{C}{T} - \frac{I^2TD}{2(K+I)}$$

Or

$$E = \frac{TDIK}{2(K+I)} + \frac{C}{T} \tag{6.10}$$

Differentiating with respect to T and equating to zero:

$$\frac{DIK}{2(K+I)} - \frac{C}{T^2} = 0$$

Or

$$\frac{DIK}{2(K+I)} = \frac{C}{T^2}$$

Or

$$T^2 = \frac{2C(K+I)}{DIK}$$

Or

$$T^* = \sqrt{\frac{2C(K+I)}{DIK}} \tag{6.11}$$

From Eq. (6.11):

$$\frac{DIK}{2(K+I)} = \frac{C}{T^2}$$

Putting this value in Eq. (6.10),

$$E^* = T\left(\frac{C}{T^2}\right) + \frac{C}{T}$$

Or

$$E^* = \frac{C}{T} + \frac{C}{T}$$

Or

$$E^* = \frac{2C}{T} \tag{6.12}$$

where T is calculated by Eq. (6.11).

Example 6.3
Consider the Example problem 6.1 with the following data:

Annual demand, D = 600 items
Ordering cost, C = ₹200
Annual inventory carrying cost per unit, I = ₹11

Additionally assume:

Annual shortage/backlogging cost per unit, K = ₹100

Now:
From Eq. (6.11), cycle time:

$$T = \sqrt{\frac{2 \times 200 \times (100 + 11)}{600 \times 11 \times 100}}$$

Or

$$T = 0.259 \text{ year}$$

With reference to the Example problem 6.1, cycle time is now longer to incorporate the stock outs.

From Eq. (6.12), the approx. total cost is:

$$E = \frac{2 \times 200}{0.259}$$

Or

$$E = ₹\ 1542.20$$

In spite of the shortages, the overall costs are lower in comparison with the basic case of Example 6.1 (i.e., where stock outs are not present).

Substituting the value of T in Eq. (6.9), optimum shortages/backlogged quantity can also be obtained, i.e.,

$$J = \frac{ITD}{(K + I)}$$

Or

$$J = \frac{11 \times 0.259 \times 600}{(100 + 11)}$$

Or

$$J = 15.4 \text{ units}$$

As discussed before, a reduction in cycle time is represented by Fig. 6.8 without affecting the input parameters.

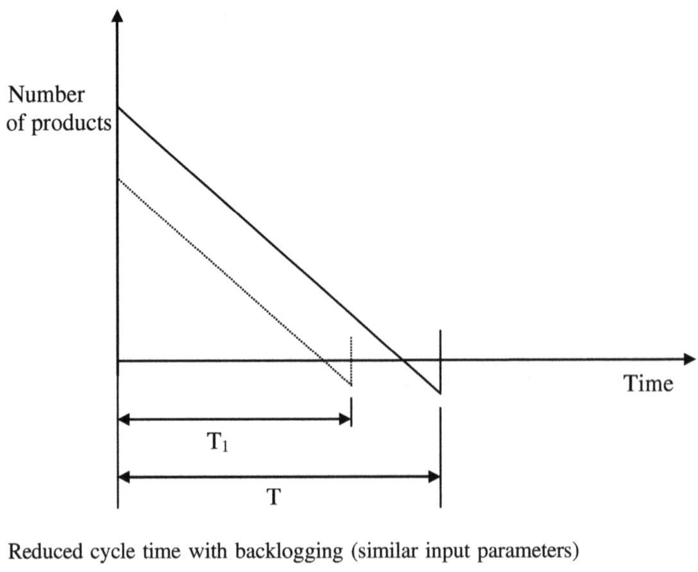

Fig. 6.8 Reduced cycle time with backlogging (similar input parameters)

6.3.1 Upward Variation in Cycle Time

Upward variation in the optimal C.T. takes effect because of the change in input parameters such as:

(a) Increase in ordering cost
(b) Decrease in demand
(c) Decrease in inventory carrying cost
(d) Decrease in stock out cost

Let:
P = % variation in parameters.
Analysis pertaining to the variation is now discussed.

(a) **Increase in ordering cost**:

$$C_1 = \left(1 + \frac{P}{100}\right)C$$

$$\begin{aligned}\text{Increase in the cycle time} &= T_1 - T \\ &= \sqrt{\frac{2C_1(K+I)}{DIK}} - \sqrt{\frac{2C(K+I)}{DIK}} \\ &= \sqrt{\frac{2C(K+I)}{DIK}}\left[\sqrt{\frac{C_1}{C}} - 1\right] \\ &= \sqrt{\frac{2C(K+I)}{DIK}}\left[\sqrt{\left(1 + \frac{P}{100}\right)} - 1\right]\end{aligned}$$

And:

$$\%\ \text{increase in C.T.} = \sqrt{\left(1 + \frac{P}{100}\right)} - 1$$

For an illustration, use the information pertaining to Example 6.3 as given below:

D	C	I	K	T	E
600	200	11	100	0.259	1542.20

After making the ordering cost higher, the parameters are given as:

% Increase in C	5%	10%	15%	20%	25%	30%
C	210	220	230	240	250	260
T	0.266	0.272	0.278	0.284	0.290	0.296
% increase in T	2.47%	4.88%	7.24%	9.54%	11.80%	14.02%
E	1580.28	1617.47	1653.82	1689.39	1724.23	1758.38
% increase in E	2.47%	4.88%	7.24%	9.54%	11.80%	14.02%

(b) **Decrease in demand**:

$$D_1 = D\left(1 - \frac{P}{100}\right)$$

$$\begin{aligned}\text{Increase in the C.T.} &= T_1 - T \\ &= \sqrt{\frac{2C(K+I)}{D_1 IK}} - \sqrt{\frac{2C(K+I)}{DIK}} \\ &= \sqrt{\frac{2C(K+I)}{DIK}}\left[\sqrt{\frac{D}{D_1}} - 1\right] \\ &= \sqrt{\frac{2C(K+I)}{DIK}}\left[\sqrt{\frac{1}{(1-\frac{P}{100})}} - 1\right]\end{aligned}$$

And:

$$\%\ \text{increase in C.T.} = \sqrt{\frac{1}{(1-\frac{P}{100})}} - 1$$

Now again:

D	C	I	K	T	E
600	200	11	100	0.259	1542.20

And:

% Decrease in D	5%	10%	15%	20%	25%	30%
D	570	540	510	480	450	420
T	0.266	0.273	0.281	0.290	0.299	0.310
% increase in T	2.60%	5.41%	8.47%	11.80%	15.47%	19.52%
E	1503.15	1463.06	1421.84	1379.38	1335.58	1290.30
% decrease in E	2.53%	5.13%	7.80%	10.56%	13.40%	16.33%

(c) **Decrease in inventory carrying cost**:

$$I_1 = I\left(1 - \frac{P}{100}\right)$$

$$\begin{aligned}
\text{Increase in the C.T.} &= T_1 - T \\
&= \sqrt{\frac{2C(K+I_1)}{DI_1K}} - \sqrt{\frac{2C(K+I)}{DIK}} \\
&= \sqrt{\frac{2C(K+I)}{DIK}}\left[\sqrt{\frac{I(K+I_1)}{I_1(K+I)}} - 1\right] \\
&= \sqrt{\frac{2C(K+I)}{DIK}}\left[\sqrt{\frac{K+I(1-P/100)}{(K+I)(1-P/100)}} - 1\right]
\end{aligned}$$

And

$$\%\ \text{increase in C.T.} = \sqrt{\frac{K+I(1-P/100)}{(K+I)(1-P/100)}} - 1$$

With the similar available information, the obtained details are as follows:

% decrease in I	5%	10%	15%	20%	25%	30%
I	10.45	9.9	9.35	8.8	8.25	7.7
T	0.265	0.272	0.279	0.287	0.296	0.305
% increase in T	2.34%	4.89%	7.66%	10.69%	14.03%	17.73%
E	1506.89	1470.36	1432.52	1393.26	1352.44	1309.92
% decrease in E	2.29%	4.66%	7.11%	9.66%	12.30%	15.06%

(d) **Decrease in stock out cost**:

$$K_1 = K\left(1 - \frac{P}{100}\right)$$

$$\begin{aligned}\text{Increase in the C.T.} &= T_1 - T\\ &= \sqrt{\frac{2C(K_1+I)}{DIK_1}} - \sqrt{\frac{2C(K+I)}{DIK}}\\ &= \sqrt{\frac{2C(K+I)}{DIK}}\left[\sqrt{\frac{K(K_1+I)}{K_1(K+I)}} - 1\right]\\ &= \sqrt{\frac{2C(K+I)}{DIK}}\left[\sqrt{\frac{I+K(1-P/100)}{(I+K)(1-P/100)}} - 1\right]\end{aligned}$$

And

$$\%\text{ increase in C.T.} = \sqrt{\frac{I+K(1-P/100)}{(I+K)(1-P/100)}} - 1$$

With the previous background:

% decrease in K	5%	10%	15%	20%	25%	30%
K	95	90	85	80	75	70
T	0.260	0.261	0.262	0.263	0.264	0.265
% increase in T	0.26%	0.55%	0.87%	1.23%	1.64%	2.10%
E	1538.19	1533.78	1528.89	1523.44	1517.34	1510.46
% decrease in E	0.26%	0.55%	0.86%	1.22%	1.61%	2.06%

The derived generalized results are also summarized in Table 6.3.

Table 6.3 Results for an upward variation in cycle time with backlog

Variation in input parameter	Variation in optimal C.T.
Increase in ordering cost	Increase in the cycle time $= \sqrt{\frac{2C(K+I)}{DIK}}\left[\sqrt{\left(1+\frac{P}{100}\right)} - 1\right]$ % increase in C.T. $= \sqrt{\left(1+\frac{P}{100}\right)} - 1$
Decrease in demand	Increase in the cycle time $= \sqrt{\frac{2C(K+I)}{DIK}}\left[\sqrt{\frac{1}{\left(1-\frac{P}{100}\right)}} - 1\right]$ % increase in C.T. $= \sqrt{\frac{1}{\left(1-\frac{P}{100}\right)}} - 1$
Decrease in inventory carrying cost	Increase in the C.T. $= \sqrt{\frac{2C(K+I)}{DIK}}\left[\sqrt{\frac{K+I(1-P/100)}{(K+I)(1-P/100)}} - 1\right]$ % increase in C.T. $= \sqrt{\frac{K+I(1-P/100)}{(K+I)(1-P/100)}} - 1$
Decrease in stock out cost	Increase in the C.T. $= \sqrt{\frac{2C(K+I)}{DIK}}\left[\sqrt{\frac{I+K(1-P/100)}{(I+K)(1-P/100)}} - 1\right]$ % increase in C.T. $= \sqrt{\frac{I+K(1-P/100)}{(I+K)(1-P/100)}} - 1$

6.3.2 Downward Variation in Cycle Time

Downward variation in an optimal C.T. is because of the following changes:

(a) Decrease in ordering cost
(b) Increase in demand
(c) Increase in inventory holding cost
(d) Increase in stock out cost

These are now discussed.

(a) **Decrease in ordering cost**:

Consider the data set:

D	C	I	K	T	E
600	200	11	100	0.259	1542.20

With the reduced ordering cost, the computational results are as follows:

% decrease in C	5%	10%	15%	20%	25%	30%
C	190	180	170	160	150	140
T	0.253	0.246	0.239	0.232	0.225	0.217
% decrease in T	2.53%	5.13%	7.80%	10.56%	13.40%	16.33%
E	1503.15	1463.06	1421.84	1379.38	1335.58	1290.30
% decrease in E	2.53%	5.13%	7.80%	10.56%	13.40%	16.33%

In order to generalize:

$$C_1 = \left(1 - \frac{P}{100}\right)C$$

$$\begin{aligned}
\text{Decrease in the cycle time} &= T - T_1 \\
&= \sqrt{\frac{2C(K+I)}{DIK}} - \sqrt{\frac{2C_1(K+I)}{DIK}} \\
&= \sqrt{\frac{2C(K+I)}{DIK}}\left[1 - \sqrt{\frac{C_1}{C}}\right] \\
&= \sqrt{\frac{2C(K+I)}{DIK}}\left[1 - \sqrt{\left(1 - \frac{P}{100}\right)}\right]
\end{aligned}$$

And:

$$\% \text{ decrease in C.T.} = 1 - \sqrt{\left(1 - \frac{P}{100}\right)}$$

(b) **Increase in demand**:

With similar reference set of data:

% increase in D	5%	10%	15%	20%	25%	30%
D	630	660	690	720	750	780
T	0.253	0.247	0.242	0.237	0.232	0.227
% decrease in T	2.41%	4.65%	6.75%	8.71%	10.56%	12.29%
E	1580.28	1617.47	1653.82	1689.39	1724.23	1758.38
% increase in E	2.47%	4.88%	7.24%	9.54%	11.80%	14.02%

For a generalization:

$$D_1 = D\left(1 + \frac{P}{100}\right)$$

$$\begin{aligned}
\text{Decrease in the C.T.} &= T - T_1 \\
&= \sqrt{\frac{2C(K+I)}{DIK}} - \sqrt{\frac{2C(K+I)}{D_1 IK}} \\
&= \sqrt{\frac{2C(K+I)}{DIK}}\left[1 - \sqrt{\frac{D}{D_1}}\right] \\
&= \sqrt{\frac{2C(K+I)}{DIK}}\left[1 - \sqrt{\frac{1}{(1 + \frac{P}{100})}}\right]
\end{aligned}$$

And:

$$\% \text{ decrease in C.T.} = 1 - \sqrt{\frac{1}{(1 + \frac{P}{100})}}$$

(c) **Increase in inventory holding cost**:

For this variation:

% increase in I	5%	10%	15%	20%	25%	30%
I	11.55	12.1	12.65	13.2	13.75	14.3
T	0.254	0.249	0.244	0.239	0.235	0.231
% decrease in T	2.17%	4.18%	6.06%	7.81%	9.46%	11.00%
E	1576.38	1609.52	1641.67	1672.90	1703.26	1732.81
% increase in E	2.22%	4.37%	6.45%	8.47%	10.44%	12.36%

Now:

$$I_1 = I\left(1 + \frac{P}{100}\right)$$

$$\begin{aligned}
\text{Decrease in the C.T.} &= T - T_1 \\
&= \sqrt{\frac{2C(K+I)}{DIK}} - \sqrt{\frac{2C(K+I_1)}{DI_1K}} \\
&= \sqrt{\frac{2C(K+I)}{DIK}}\left[1 - \sqrt{\frac{I(K+I_1)}{I_1(K+I)}}\right] \\
&= \sqrt{\frac{2C(K+I)}{DIK}}\left[1 - \sqrt{\frac{K+I(1+P/100)}{(K+I)(1+P/100)}}\right]
\end{aligned}$$

And:

$$\%\ \text{decrease in C.T.} = 1 - \sqrt{\frac{K+I(1+P/100)}{(K+I)(1+P/100)}}$$

(d) **Increase in stock out cost**:

With this variation:

% increase in K	5%	10%	15%	20%	25%	30%
K	105	110	115	120	125	130
T	0.259	0.258	0.258	0.257	0.257	0.256
% decrease in T	0.24%	0.45%	0.65%	0.83%	1.00%	1.15%
E	1545.85	1549.19	1552.26	1555.10	1557.71	1560.14
% increase in E	0.24%	0.45%	0.65%	0.84%	1.01%	1.16%

Table 6.4 Results for a downward variation in cycle time with backlog

Variation in input parameter	Variation in optimal C.T.
Decrease in ordering cost	Decrease in the cycle time = $\sqrt{\frac{2C(K+I)}{DIK}}\left[1-\sqrt{\left(1-\frac{P}{100}\right)}\right]$ % decrease in C.T. = $1-\sqrt{\left(1-\frac{P}{100}\right)}$
Increase in demand	Decrease in the cycle time = $\sqrt{\frac{2C(K+I)}{DIK}}\left[1-\sqrt{\frac{1}{\left(1+\frac{P}{100}\right)}}\right]$ % decrease in C.T. = $1-\sqrt{\frac{1}{\left(1+\frac{P}{100}\right)}}$
Increase in inventory holding cost	Decrease in the C.T. = $\sqrt{\frac{2C(K+I)}{DIK}}\left[1-\sqrt{\frac{K+I(1+P/100)}{(K+I)(1+P/100)}}\right]$ % decrease in C.T. = $1-\sqrt{\frac{K+I(1+P/100)}{(K+I)(1+P/100)}}$
Increase in stock out cost	Decrease in the C.T. = $\sqrt{\frac{2C(K+I)}{DIK}}\left[1-\sqrt{\frac{I+K(1+P/100)}{(I+K)(1+P/100)}}\right]$ % decrease in C.T. = $1-\sqrt{\frac{I+K(1+P/100)}{(I+K)(1+P/100)}}$

In order to get the generalized result:

$$K_1 = K\left(1+\frac{P}{100}\right)$$

$$\begin{aligned}
\text{Decrease in the C.T.} &= T - T_1 \\
&= \sqrt{\frac{2C(K+I)}{DIK}} - \sqrt{\frac{2C(K_1+I)}{DIK_1}} \\
&= \sqrt{\frac{2C(K+I)}{DIK}}\left[1-\sqrt{\frac{K(K_1+I)}{K_1(K+I)}}\right] \\
&= \sqrt{\frac{2C(K+I)}{DIK}}\left[1-\sqrt{\frac{I+K(1+P/100)}{(I+K)(1+P/100)}}\right]
\end{aligned}$$

And

$$\%\ \text{decrease in C.T.} = 1-\sqrt{\frac{I+K(1+P/100)}{(I+K)(1+P/100)}}$$

A generalization has been made in arriving at the results, and these are also shown in Table 6.4.

6.4 Interaction of Parameters Without Stock Out

In context to the cycle time and its variation, different parameters play their role. Cycle time varies when a parameter changes. However, either because of internal situation or buyer requirement, it may be desirable to keep similar cycle time. This might be achieved by understanding the interaction of relevant parameters.

6.4.1 Decreased Cycle Time

Optimal cycle time decreases because of higher demand and lower ordering cost among others.

(a) **Higher demand**

Consider the data as follows:

D	C	I	T	E
600	200	11	0.246	1624.81

Now:

$$T = \sqrt{\frac{2C}{DI}}$$

With 10% increase in demand, reduced cycle time = 0.235 year. In order to have similar cycle time as before, option such as variation in I might be available.

Decrease in I:
Reduced value of holding cost (I_1) is obtained as follows:

$$\sqrt{\frac{2 \times 200}{660 I_1}} = 0.246$$

Or

$$I_1 = 10.01$$

For a general approach, let:
P = % increase in demand
M = % decrease in holding cost
Therefore:

$$D_1 = \left(1 + \frac{P}{100}\right) D$$

$$I_1 = \left(1 - \frac{M}{100}\right) I$$

For similar cycle time:

$$\sqrt{\frac{2C}{D_1 I_1}} = \sqrt{\frac{2C}{DI}}$$

Or

$$D_1 I_1 = DI$$

Or

$$\left(1 + \frac{P}{100}\right)\left(1 - \frac{M}{100}\right) = 1$$

Or

$$M = \frac{P}{(1 + P/100)}$$

Variation in M with respect to P is shown in Table 6.5. Values of M are lower than the respective values of P. Furthermore M is less sensitive with respect to the higher values of P.

(b) **Lower ordering cost**
With basic data set as before, implement 10% reduction in ordering cost, i.e.:

$$C_1 = ₹180$$

And the reduced cycle time = 0.234 year.
In order to have similar cycle time as before, option such as variation in I might be available.

Decrease in I:
Reduced value of holding cost (I_1) is obtained as follows:

$$\sqrt{\frac{2 \times 180}{600 I_1}} = 0.246$$

Or

$$I_1 = 9.9$$

Table 6.5 Variation in M with respect to P for decreased I (without) stock out

S. No.	P	$M = \frac{P}{(1+P/100)}$
1	10	9.09
2	20	16.67
3	30	23.08
4	40	28.57
5	50	33.33

For a general approach, let:
P = % decrease in ordering cost
M = % decrease in holding cost
Therefore:

$$C_1 = \left(1 - \frac{P}{100}\right)C$$

$$I_1 = \left(1 - \frac{M}{100}\right)I$$

For similar cycle time:

$$\sqrt{\frac{2C_1}{DI_1}} = \sqrt{\frac{2C}{DI}}$$

Or

$$\frac{C_1}{I_1} = \frac{C}{I}$$

Or

$$\left(1 - \frac{P}{100}\right) = \left(1 - \frac{M}{100}\right)$$

Or

$$M = P$$

6.4.2 *Higher Cycle Time*

Optimal cycle time increases because of:

(a) Lower inventory holding cost
(b) Lower demand
(c) Higher ordering cost

Consider the reference data as follows for the above factors:

D	C	I	T	E
600	200	11	0.246	1624.81

(a) **Lower inventory holding cost**
Now:

$$T = \sqrt{\frac{2C}{DI}}$$

With 10% reduction in *I*, increased cycle time = 0.259 year. In order to have similar cycle time as before, option such as variation in C and D might be available.

(i) Decrease in C:

Reduced value of ordering cost (C_1) is obtained as follows:

$$\sqrt{\frac{2 \times C_1}{600 \times 9.9}} = 0.246$$

Or

$$C_1 = ₹\ 180$$

For a general approach, let:
P = % reduction in holding cost
M = % reduction in ordering cost
Therefore:

$$I_1 = \left(1 - \frac{P}{100}\right) I$$

$$C_1 = \left(1 - \frac{M}{100}\right) C$$

For similar cycle time:

$$\sqrt{\frac{2C_1}{DI_1}} = \sqrt{\frac{2C}{DI}}$$

Or

$$\frac{C_1}{I_1} = \frac{C}{I}$$

Or

$$\left(1 - \frac{M}{100}\right) = \left(1 - \frac{P}{100}\right)$$

Or

$$M = P$$

(ii) Increase in D:

If it is possible, a potential increase in demand may be explored. An increased value of demand (D_1) is obtained as follows:

$$\sqrt{\frac{2 \times 200}{9.9 D_1}} = 0.246$$

Or

$$D_1 = 667$$

For a general approach, let:
P = % reduction in holding cost
M = % increase in demand
Therefore:

$$I_1 = \left(1 - \frac{P}{100}\right) I$$

$$D_1 = \left(1 + \frac{M}{100}\right) D$$

For similar cycle time:

$$\sqrt{\frac{2C}{D_1 I_1}} = \sqrt{\frac{2C}{DI}}$$

Or

$$D_1 I_1 = DI$$

Or

$$\left(1 + \frac{M}{100}\right)\left(1 - \frac{P}{100}\right) = 1$$

Or

$$M = \frac{P}{(1 - P/100)}$$

Table 6.6 Variation in M with respect to P for increased D (without) stock out

S. No.	P	$M = \frac{P}{(1-P/100)}$
1	10	11.11
2	20	25
3	30	42.86
4	40	65.67
5	50	100

Variation in M with respect to P is shown in Table 6.6. Values of M are higher than the respective values of P. Furthermore, M is more sensitive with respect to the higher values of P.

(b) **Lower demand**

In case of lower demand, a relevant option may be to reduce the ordering cost.

$$\sqrt{\frac{2C_1}{D_1 I}} = \sqrt{\frac{2C}{DI}}$$

Or

$$\frac{C_1}{D_1} = \frac{C}{D}$$

Or

$$\left(1 - \frac{M}{100}\right) = \left(1 - \frac{P}{100}\right)$$

Or

$$M = P$$

where:
P = % reduction in demand
M = % decrease in ordering cost

(c) **Higher ordering cost**

When ordering cost increases, a relevant option might be to increase the demand.

$$\sqrt{\frac{2C_1}{D_1 I}} = \sqrt{\frac{2C}{DI}}$$

Or

$$\frac{C_1}{D_1} = \frac{C}{D}$$

Or

$$\left(1+\frac{M}{100}\right)=\left(1+\frac{P}{100}\right)$$

Or

$$M=P$$

where:

P = % increase in ordering cost
M = % increase in demand

6.5 Interaction of Parameters with Stock Out

In order to keep similar cycle time, interaction of parameters is now discussed.

6.5.1 *Increased Cycle Time*

Cycle time increases because of the following changes:

(a) Increase in ordering cost
(b) Decrease in demand
(c) Decrease in inventory carrying cost
(d) Decrease in stock out cost

Let:

P = % variation in these parameters
M = % variation in other relevant parameters in order to have similar cycle time as before.

Analysis pertaining to the variation is now discussed.

(a) **Increase in ordering cost**

For similar cycle time, a relevant option is potential increase in demand. From Eq. (6.11), optimal C.T.:

$$T=\sqrt{\frac{2C(K+I)}{DIK}}$$

The related information is as follows:

D	C	I	K	T	E
600	200	11	100	0.259	1542.20

After 20% increase in ordering cost, an increased demand D_1 for similar cycle time is obtained as follows:

$$\sqrt{\frac{2 \times 240 \times (100+11)}{11 \times 100 D_1}} = \sqrt{\frac{2 \times 200 \times (100+11)}{600 \times 11 \times 100}}$$

Or

$$D_1 = 720$$

For a general approach, C_1 and D_1 are varied parameters as follows:

$$C_1 = \left(1 + \frac{P}{100}\right) C$$

$$D_1 = \left(1 + \frac{M}{100}\right) D$$

Now, for similar cycle time:

$$\sqrt{\frac{2C_1(K+I)}{D_1 IK}} = \sqrt{\frac{2C(K+I)}{DIK}}$$

Or

$$\frac{C_1}{D_1} = \frac{C}{D}$$

Or

$$\left(1 + \frac{M}{100}\right) = \left(1 + \frac{P}{100}\right)$$

Or

$$M = P$$

(b) **Decrease in demand**

It can be handled by a decrease in ordering cost, such as:

$$\sqrt{\frac{2C_1(K+I)}{D_1IK}} = \sqrt{\frac{2C(K+I)}{DIK}}$$

Or

$$\frac{C_1}{D_1} = \frac{C}{D}$$

Or

$$\left(1 - \frac{M}{100}\right) = \left(1 - \frac{P}{100}\right)$$

Or

$$M = P$$

(c) **Decrease in inventory carrying cost**

For a similar cycle time, the following changes might be suitable:

(i) Increase in demand
(ii) Decrease in ordering cost

Now these are analyzed.

(i) Increase in demand:
The basic information is as follows:

D	C	I	K	T	E
600	200	11	100	0.259	1542.20

After 20% decrease in carrying cost, an increased demand D_1 for similar cycle time is obtained as follows:

$$\sqrt{\frac{2 \times 200 \times (100 + 8.8)}{8.8 \times 100D_1}} = \sqrt{\frac{2 \times 200 \times (100 + 11)}{600 \times 11 \times 100}}$$

Or

$$D_1 = 735$$

For a generalization:

$$\sqrt{\frac{2C(K+I_1)}{D_1 I_1 K}} = \sqrt{\frac{2C(K+I)}{DIK}}$$

Or

$$D_1 I_1 (K+I) = DI(K+I_1)$$

Or

$$\left(1+\frac{M}{100}\right)\left(1-\frac{P}{100}\right)(K+I) = K+I\left(1-\frac{P}{100}\right)$$

Or

$$1+\frac{M}{100} = \frac{K+I(1-P/100)}{(1-P/100)(K+I)}$$

Or

$$\frac{M}{100} = \frac{K+I(1-P/100)}{(1-P/100)(K+I)} - 1$$

Or

$$\frac{M}{100} = \frac{K+I(1-P/100)-(1-P/100)(K+I)}{(1-P/100)(K+I)}$$

Or

$$\frac{M}{100} = \frac{K-(1-P/100)K}{(1-P/100)(K+I)}$$

Or

$$\frac{M}{100} = \frac{(KP/100)}{(1-P/100)(K+I)}$$

Or

$$M = \frac{KP}{(1-P/100)(K+I)}$$

Table 6.7 Variation in M (increased D) with respect to P (decreased I) with stock out

S. No.	P	$M = \frac{KP}{(1-P/100)(K+I)}$
1	10	10.01
2	20	22.52
3	30	38.61
4	40	60.06
5	50	90.09

For the current information, variation in M is shown in Table 6.7 with respect to P. Values of M for the present situation are higher in comparison with P, and the sensitivity of M increases with the higher values of P.
While comparing with the results of Table 6.6, the values of M are lower. This is because:

$$\frac{KP}{(1 - P/100)(K + I)} < \frac{P}{(1 - P/100)}$$

(ii) Decrease in ordering cost:

The basic information is as follows:

D	C	I	K	T	E
600	200	11	100	0.259	1542.20

After 20% decrease in carrying cost, a decreased ordering cost C_1 for similar cycle time is obtained as follows:

$$\sqrt{\frac{2 \times C_1 \times (100 + 8.8)}{600 \times 8.8 \times 100}} = \sqrt{\frac{2 \times 200 \times (100 + 11)}{600 \times 11 \times 100}}$$

Or

$$C_1 = ₹\,163.24$$

For a generalization:

$$\sqrt{\frac{2C_1(K + I_1)}{DI_1K}} = \sqrt{\frac{2C(K + I)}{DIK}}$$

Or

$$\frac{C_1(K + I_1)}{I_1} = \frac{C(K + I)}{I}$$

Or

$$IC_1(K + I_1) = CI_1(K + I)$$

Or

$$\left(1 - \frac{M}{100}\right)\left[K + I\left(1 - \frac{P}{100}\right)\right] = \left(1 - \frac{P}{100}\right)(K + I)$$

Or

$$1 - \frac{M}{100} = \frac{(K + I)(1 - P/100)}{K + I(1 - P/100)}$$

Or

$$\frac{M}{100} = 1 - \frac{(K + I)(1 - P/100)}{K + I(1 - P/100)}$$

Or

$$\frac{M}{100} = \frac{K + I(1 - P/100) - (K + I)(1 - P/100)}{K + I(1 - P/100)}$$

Or

$$\frac{M}{100} = \frac{K - K(1 - P/100)}{K + I(1 - P/100)}$$

Or

$$\frac{M}{100} = \frac{(KP/100)}{K + I(1 - P/100)}$$

Or

$$M = \frac{KP}{K + I(1 - P/100)}$$

For the present information, variation in M is shown in Table 6.8 with respect to P. Values of M for the present situation are lower in comparison with P, and the sensitivity of M increases with higher values of P.

Table 6.8 Variation in M (reduced C) with respect to P (decreased I) with stock out

S. No.	P	$M = \frac{KP}{K+I(1-P/100)}$
1	10	9.099
2	20	18.38
3	30	27.86
4	40	37.52
5	50	47.39

While comparing with Table 6.7, the values of M are lower. This is because:

$$\frac{KP}{K+I(1-P/100)} < \frac{KP}{(1-P/100)(K+I)}$$

Or

$$K+I(1-P/100) > (1-P/100)(K+I)$$

Or

$$K > (1-P/100)K$$

And that is true.

While increasing the demand, additional profit is also gained. However, when it is difficult to increase the demand, an effort may be made to reduce the ordering cost if it is possible.

(d) **Decrease in stock out cost**

For a similar cycle time, the relevant options might be: (i) Increase in demand (ii) Reduction in ordering cost

(i) Increase in demand:

The present details are as follows:

D	C	I	K	T	E
600	200	11	100	0.259	1542.20

After 20% decrease in K, an increased demand D_1 for similar cycle time is obtained as follows:

$$\sqrt{\frac{2\times 200\times(80+11)}{11\times 80D_1}} = \sqrt{\frac{2\times 200\times(100+11)}{600\times 11\times 100}}$$

Or

$$D_1 = 615$$

For a generalization:

$$\sqrt{\frac{2C(K_1+I)}{D_1 I K_1}} = \sqrt{\frac{2C(K+I)}{DIK}}$$

Or

$$D_1 K_1 (K+I) = DK(K_1+I)$$

Or

$$\left(1+\frac{M}{100}\right)\left(1-\frac{P}{100}\right)(K+I) = K\left(1-\frac{P}{100}\right)+I$$

Or

$$1+\frac{M}{100} = \frac{I+K(1-P/100)}{(1-P/100)(K+I)}$$

Or

$$\frac{M}{100} = \frac{I+K(1-P/100)}{(1-P/100)(K+I)} - 1$$

Or

$$\frac{M}{100} = \frac{I+K(1-P/100)-(1-P/100)(K+I)}{(1-P/100)(K+I)}$$

Or

$$\frac{M}{100} = \frac{I-(1-P/100)I}{(1-P/100)(K+I)}$$

Or

$$\frac{M}{100} = \frac{(IP/100)}{(1-P/100)(K+I)}$$

Table 6.9 Variation in M for increased D (with stock out cost reduction)

S. No.	P	$M = \frac{IP}{(1-P/100)(K+I)}$
1	10	1.10
2	20	2.48
3	30	4.25
4	40	6.61
5	50	9.91

Or

$$M = \frac{IP}{(1 - P/100)(K+I)}$$

For the current information, variation in M is shown in Table 6.9 with respect to P. Values of M for the present situation are much lower in comparison with P, and the sensitivity of M increases with higher values of P.

(ii) Reduction in ordering cost:

The present details are as follows:

D	C	I	K	T	E
600	200	11	100	0.259	1542.20

After 20% decrease in stock out cost, a reduced ordering cost C_1 for similar cycle time is obtained as follows:

$$\sqrt{\frac{2 \times C_1 \times (80+11)}{600 \times 11 \times 80}} = \sqrt{\frac{2 \times 200 \times (100+11)}{600 \times 11 \times 100}}$$

Or

$$C_1 = ₹\ 195.16$$

For a generalization:

$$\sqrt{\frac{2C_1(K_1+I)}{DIK_1}} = \sqrt{\frac{2C(K+I)}{DIK}}$$

Or

$$\frac{C_1(K_1+I)}{K_1} = \frac{C(K+I)}{K}$$

Or

$$KC_1(K_1 + I) = CK_1(K + I)$$

Or

$$\left(1 - \frac{M}{100}\right)\left[K\left(1 - \frac{P}{100}\right) + I\right] = \left(1 - \frac{P}{100}\right)(K + I)$$

Or

$$1 - \frac{M}{100} = \frac{(K + I)(1 - P/100)}{I + K(1 - P/100)}$$

Or

$$\frac{M}{100} = 1 - \frac{(K + I)(1 - P/100)}{I + K(1 - P/100)}$$

Or

$$\frac{M}{100} = \frac{I + K(1 - P/100) - (K + I)(1 - P/100)}{I + K(1 - P/100)}$$

Or

$$\frac{M}{100} = \frac{I - I(1 - P/100)}{I + K(1 - P/100)}$$

Or

$$\frac{M}{100} = \frac{(IP/100)}{I + K(1 - P/100)}$$

Or

$$M = \frac{IP}{I + K(1 - P/100)}$$

For the present information, variation in M is shown in Table 6.10 with respect to P. Values of M for the present situation are much lower in comparison with P, and the sensitivity of M increases with higher values of P.

While comparing with Table 6.9, values of M are slightly lower. This is because:

Table 6.10 Variation in M for reduced C (with stock out cost reduction)

S. No.	P	$M = \frac{IP}{I+K(1-P/100)}$
1	10	1.09
2	20	2.42
3	30	4.07
4	40	6.20
5	50	9.02

$$\frac{IP}{I+K(1-P/100)} < \frac{IP}{(1-P/100)(K+I)}$$

Or

$$I+K(1-P/100) > (1-P/100)(K+I)$$

Or

$$I > (1-P/100)I$$

And that is true.

6.5.2 *Lower Cycle Time*

Cycle time reduces because of the following changes:

(i) Reduction in ordering cost
(ii) Increase in demand
(iiI) Increase in inventory holding cost
(iv) Increase in stock out cost

Let:

P = % variation in these parameters
M = % variation in other relevant parameters in order to have similar cycle time

Analysis pertaining to the variation is now discussed.

(i) **Reduction in ordering cost**

For similar cycle time, the following options might be exercised:

(a) Reduced holding cost
(b) Reduced stock out cost

From Eq. (6.11), optimal C.T.:

$$T = \sqrt{\frac{2C(K+I)}{DIK}}$$

The available data are as follows:

D	C	I	K	T	E
600	200	11	100	0.259	1542.20

(a) Reduced holding cost:

After 10% reduction in ordering cost, a reduced holding cost I_1 for similar cycle time is obtained as follows:

$$\sqrt{\frac{2 \times 180 \times (100 + I_1)}{600 \times 100 I_1}} = \sqrt{\frac{2 \times 200 \times (100 + 11)}{600 \times 11 \times 100}}$$

Or

$$22200 I_1 = 198000 + 1980 I_1$$

Or

$$20220 I_1 = 198000$$

Or

$$I_1 = 9.79$$

For a generalization:

$$\sqrt{\frac{2C_1(K+I_1)}{DI_1K}} = \sqrt{\frac{2C(K+I)}{DIK}}$$

Or

$$\frac{C_1(K+I_1)}{I_1} = \frac{C(K+I)}{I}$$

Or

$$IC_1(K+I_1) = CI_1(K+I)$$

Or

$$I_1[C(K+I) - IC_1] = IC_1K$$

Or

$$I_1 = \frac{IC_1K}{C(K+I) - IC_1}$$

Or

$$I\left(1 - \frac{M}{100}\right) = \frac{IKC(1 - P/100)}{C(K+I) - IC(1 - P/100)}$$

Or

$$1 - \frac{M}{100} = \frac{KC(1 - P/100)}{CK + (ICP/100)}$$

Or

$$1 - \frac{M}{100} = \frac{K(1 - P/100)}{K + (IP/100)}$$

Or

$$\frac{M}{100} = 1 - \frac{K(1 - P/100)}{K + (IP/100)}$$

Or

$$\frac{M}{100} = \frac{K + (IP/100) - K(1 - P/100)}{K + (IP/100)}$$

Or

$$\frac{M}{100} = \frac{(IP/100) + (KP/100)}{K + (IP/100)}$$

Table 6.11 Variation in M (reduced I) corresponding to P (reduced C) with stock out

S. No.	P	$M = \frac{P(I+K)}{K+(IP/100)}$
1	10	10.98
2	20	21.72
3	30	32.24
4	40	42.53
5	50	52.61

Or

$$M = \frac{P(I+K)}{K+(IP/100)}$$

For the current information, variation in M is shown in Table 6.11 with respect to P. Values of M for the present situation are higher in comparison with P, and the sensitivity of M decreases with higher values of P in this case.

(b) Reduced stock out cost

After 10% reduction in ordering cost, a reduced stock out cost K_1 for similar cycle time is obtained as follows:

$$\sqrt{\frac{2 \times 180 \times (K_1+11)}{600 \times 11K_1}} = \sqrt{\frac{2 \times 200 \times (100+11)}{600 \times 11 \times 100}}$$

Or

$$22200K_1 = 18000K_1 + 198000$$

Or

$$4200K_1 = 198000$$

Or

$$K_1 = 47.14$$

For a generalization:

$$\sqrt{\frac{2C_1(K_1+I)}{DIK_1}} = \sqrt{\frac{2C(K+I)}{DIK}}$$

Or

$$\frac{C_1(K_1+I)}{K_1} = \frac{C(K+I)}{K}$$

Or

$$KC_1(K_1 + I) = CK_1(K + I)$$

Or

$$K_1[C(K + I) - KC_1] = IC_1K$$

Or

$$K_1 = \frac{IC_1K}{C(K + I) - KC_1}$$

Or

$$K\left(1 - \frac{M}{100}\right) = \frac{IKC(1 - P/100)}{C(K + I) - KC(1 - P/100)}$$

Or

$$1 - \frac{M}{100} = \frac{IC(1 - P/100)}{CI + (KCP/100)}$$

Or

$$1 - \frac{M}{100} = \frac{I(1 - P/100)}{I + (KP/100)}$$

Or

$$\frac{M}{100} = 1 - \frac{I(1 - P/100)}{I + (KP/100)}$$

Or

$$\frac{M}{100} = \frac{I + (KP/100) - I(1 - P/100)}{I + (KP/100)}$$

Or

$$\frac{M}{100} = \frac{(KP/100) + (IP/100)}{I + (KP/100)}$$

Table 6.12 Variation in M (reduced K) with respect to P (reduced C)

S. No.	P	$M = \frac{P(K+I)}{I+(KP/100)}$
1	10	52.86
2	20	71.61
3	30	81.22
4	40	87.06
5	50	90.98

Or

$$M = \frac{P(K+I)}{I+(KP/100)}$$

For the current data, variation in M is shown in Table 6.12 with respect to P. Values of M for the present situation are much higher in comparison with P, and the sensitivity of M decreases with higher values of P in this case.

Comparing with the results of Table 6.11, the values of M are higher. This will happen when:

$$\frac{P(K+I)}{I+(KP/100)} > \frac{P(K+I)}{K+(IP/100)}$$

Or

$$I+(KP/100) < K+(IP/100)$$

Or

$$I(1-P/100) < K(1-P/100)$$

Or

$$I < K$$

In most of the practical situations, this might be true. Under such operational parameters, the option of a reduction in holding cost is preferred if it is convenient to do so.

(ii) **Increase in demand**

Because of an increase in demand, cycle time is lowered. For a similar cycle time as before, the following options are explored:

(a) Reduction in holding cost
(b) Reduction in stock out cost

An available set of information is as follows:

D	C	I	K	T	E
600	200	11	100	0.259	1542.20

Now a general approach is presented for each option along with the example.

(a) Reduction in holding cost:

With 20% increase in demand:

$$\sqrt{\frac{2 \times 200 \times (100 + I_1)}{720 \times 100 I_1}} = \sqrt{\frac{2 \times 200 \times (100 + 11)}{600 \times 11 \times 100}}$$

Or

$$79920 I_1 = 660000 + 6600 I_1$$

Or

$$73320 I_1 = 660000$$

Or

$$I_1 = 9$$

For a general approach:

$$\sqrt{\frac{2C(K + I_1)}{D_1 I_1 K}} = \sqrt{\frac{2C(K + I)}{DIK}}$$

Or

$$D_1 I_1 (K + I) = DI(K + I_1)$$

Or

$$D_1 I_1 (K + I) - DII_1 = DIK$$

Or

$$I_1 [D_1 (K + I) - DI] = DIK$$

Or

$$I_1 = \frac{DIK}{D_1(K+I) - DI}$$

Or

$$1 - \frac{M}{100} = \frac{K}{(1+P/100)(K+I) - I}$$

Or

$$\frac{M}{100} = 1 - \frac{K}{(1+P/100)(K+I) - I}$$

Or

$$\frac{M}{100} = \frac{(1+P/100)(K+I) - (I+K)}{(1+P/100)(K+I) - I}$$

Or

$$\frac{M}{100} = \frac{(K+I)(P/100)}{(K+I) + (K+I)(P/100) - I}$$

Or

$$M = \frac{(K+I)P}{K + (K+I)(P/100)}$$

In order to demonstrate the variation in M with respect to P, make use of the following available information:

$$K = 100$$
$$I = 11$$

Variation in M is shown in Table 6.13 with respect to P. Values of M for the present situation are lower in comparison with P, and the sensitivity of M decreases with higher values of P in this case.

Table 6.13 Variation in M (reduced I) corresponding to P (increased D) with stock out

S. No.	P	$M = \frac{(K+I)P}{K+(K+I)(P/100)}$
1	10	9.99
2	20	18.17
3	30	24.98
4	40	30.75
5	50	35.69

(b) Reduction in stock out cost:

With 20% increase in demand:

$$\sqrt{\frac{2 \times 200 \times (K_1 + 11)}{720 \times 11K_1}} = \sqrt{\frac{2 \times 200 \times (100 + 11)}{600 \times 11 \times 100}}$$

Or

$$79920K_1 = 60000K_1 + 660000$$

Or

$$19920I_1 = 660000$$

Or

$$K_1 = 33.13$$

For a general approach:

$$\sqrt{\frac{2C(K_1 + I)}{D_1 IK_1}} = \sqrt{\frac{2C(K + I)}{DIK}}$$

Or

$$D_1K_1(K + I) = DK(K_1 + I)$$

Or

$$D_1K_1(K + I) - DKK_1 = DKI$$

Or

$$K_1[D_1(K + I) - DK] = DKI$$

Or

$$K_1 = \frac{DKI}{D_1(K + I) - DK}$$

Or

$$1 - \frac{M}{100} = \frac{I}{(1 + P/100)(K + I) - K}$$

Or

$$\frac{M}{100} = 1 - \frac{I}{(1 + P/100)(K + I) - K}$$

Or

$$\frac{M}{100} = \frac{(1 + P/100)(K + I) - (K + I)}{(1 + P/100)(K + I) - K}$$

Or

$$\frac{M}{100} = \frac{(K + I)(P/100)}{(K + I) + (K + I)(P/100) - K}$$

Or

$$M = \frac{(K + I)P}{I + (K + I)(P/100)}$$

In order to demonstrate the variation in M with respect to P, make use of the following available information:

$$K = 100$$
$$I = 11$$

Variation in M is shown in Table 6.14 with respect to P. Values of M for the present situation are much higher in comparison with P, and the sensitivity of M decreases with higher values of P in this case.

Comparing with the Table 6.13, the values are much higher. This is because:

$$\frac{(K + I)P}{I + (K + I)(P/100)} > \frac{(K + I)P}{K + (K + I)(P/100)}$$

Or

$$I < K$$

Table 6.14 Variation in M (reduced K) corresponding to P (increased D)

S. No.	P	$M = \frac{(K+I)P}{I + (K+I)(P/100)}$
1	10	50.23
2	20	66.87
3	30	75.17
4	40	80.14
5	50	83.46

As mentioned before, this might be true in most of the practical situations. With such operational parameters, the option of a reduction in holding cost is preferred if it is convenient to do so.

(iii) **Increase in inventory holding cost**

For the similar cycle time, a relevant option might be a reduction in stock out cost. Now:

D	*C*	*I*	*K*	*T*	*E*
600	200	11	100	0.259	1542.20

With 10% increase in *I*:

$$\sqrt{\frac{2 \times 200 \times (K_1 + 12.1)}{600 \times 12.1 K_1}} = \sqrt{\frac{2 \times 200 \times (100 + 11)}{600 \times 11 \times 100}}$$

Or

$$1343.1 K_1 = 1100 K_1 + 13310$$

Or

$$243.1 K_1 = 13310$$

Or

$$K_1 = 54.75$$

For a generalized approach:

$$\sqrt{\frac{2C(K_1 + I_1)}{DI_1K_1}} = \sqrt{\frac{2C(K + I)}{DIK}}$$

Or

$$I_1K_1(K + I) = IK(K_1 + I_1)$$

Or

$$I_1K_1(K + I) - IKK_1 = IKI_1$$

Or

$$K_1[I_1(K + I) - IK] = IKI_1$$

Or

$$K_1 = \frac{IKI_1}{I_1(K+I) - IK}$$

Or

$$1 - \frac{M}{100} = \frac{II_1}{I_1(K+I) - IK}$$

Or

$$\frac{M}{100} = 1 - \frac{II_1}{I_1(K+I) - IK}$$

Or

$$\frac{M}{100} = \frac{I_1(K+I) - IK - II_1}{I_1(K+I) - IK}$$

Or

$$\frac{M}{100} = \frac{I_1K - IK}{I_1K + II_1 - IK}$$

Or

$$\frac{M}{100} = \frac{K(I_1 - I)}{K(I_1 - I) + II_1}$$

Or

$$\frac{M}{100} = \frac{IK[(1 + P/100) - 1]}{KI[(1 + P/100) - 1] + II_1}$$

Or

$$\frac{M}{100} = \frac{(KP/100)}{(KP/100) + I(1 + P/100)}$$

Or

$$M = \frac{KP}{I + (K+I)(P/100)}$$

Table 6.15 Variation in M (reduced K) corresponding to P (increased I)

S. No.	P	$M = \frac{KP}{I+(K+I)(P/100)}$
1	10	45.25
2	20	60.24
3	30	67.72
4	40	72.20
5	50	75.19

In order to demonstrate the variation in M with respect to P, consider the available data:

$$K = 100$$
$$I = 11$$

Variation in M is shown in Table 6.15 with respect to P. Values of M for the present situation are higher in comparison with P, and the sensitivity of M decreases with higher values of P.

(iv) **Increase in stock out cost**

A reduction in holding cost might be a suitable option for keeping similar cycle time. Now:

D	C	I	K	T	E
600	200	11	100	0.259	1542.20

With 50% increase in K:

$$\sqrt{\frac{2 \times 200 \times (150 + I_1)}{600 \times 150 I_1}} = \sqrt{\frac{2 \times 200 \times (100 + 11)}{600 \times 11 \times 100}}$$

Or

$$16650 I_1 = 165000 + 1100 I_1$$

Or

$$15550 I_1 = 165000$$

Or

$$I_1 = 10.61$$

For a generalized approach:

$$\sqrt{\frac{2C(K_1 + I_1)}{D I_1 K_1}} = \sqrt{\frac{2C(K + I)}{DIK}}$$

Or

$$I_1K_1(K+I) = IK(K_1+I_1)$$

Or

$$I_1K_1(K+I) - I_1IK = IKK_1$$

Or

$$I_1[K_1(K+I) - IK] = IKK_1$$

Or

$$I_1 = \frac{IKK_1}{K_1(K+I) - IK}$$

Or

$$1 - \frac{M}{100} = \frac{KK_1}{K_1(K+I) - IK}$$

Or

$$\frac{M}{100} = 1 - \frac{KK_1}{K_1(K+I) - IK}$$

Or

$$\frac{M}{100} = \frac{K_1(K+I) - IK - KK_1}{K_1(K+I) - IK}$$

Or

$$\frac{M}{100} = \frac{I(K_1 - K)}{I(K_1 - K) + KK_1}$$

Or

$$\frac{M}{100} = \frac{IK[(1+P/100) - 1]}{KI[(1+P/100) - 1] + KK(1+P/100)}$$

Table 6.16 Variation in M (reduced I) corresponding to P (increased K)

S. No.	P	$M = \frac{IP}{K+(K+I)(P/100)}$
1	10	0.99
2	20	1.8
3	30	2.48
4	40	3.05
5	50	3.54

Or

$$\frac{M}{100} = \frac{(IP/100)}{(IP/100) + K(1 + P/100)}$$

Or

$$M = \frac{IP}{K + (K + I)(P/100)}$$

In order to demonstrate the variation in M with respect to P, consider the following data:

$$K = 100$$
$$I = 11$$

Variation in M is shown in Table 6.16 with respect to P. Values of M for the present situation are much lower in comparison with P, and the sensitivity of M decreases with higher values of P.

Chapter 7
Stock Out

As long as there is availability of a product for sale, there is no stock out/shortage. However, when there is demand of a product, but it is not available; then, the stock out/shortage situation is being faced by the retailer/wholesaler for that product. For an analysis of such case, stock out cost needs to be estimated.

7.1 Stock Out Cost Estimation

When shortages are faced, it may not affect only the customer, but also the retailer. For satisfying the customer, the retailer has to make extra efforts and these efforts may be included in the assessment of the cost of stock out. Suppose that some of the customers have stopped visiting a retailer or purchasing a particular brand of a product, then there might be a need for advertising related to a particular product brand/company. Therefore, the effects of shortages may not confine to only the customer/retailer. But it may extend to the wholesaler and producer company also. For instance, when an item is in shortage, the retailer may demand that item urgently from the wholesaler and there is a need for urgent transportation of that item resulting into an additional cost. This affects the operations of wholesaler as well as retailer since it was not planned earlier. A different transportation medium may also be used which is not so economical.

In order to estimate the stock out cost, an attention to the following aspects is useful:

(a) In case where a product is not available at the retailer location, an idea for loss of profit is helpful. This loss of profit relates to the situation when demand is available, i.e., a customer is present asking for that product. But the customer is unable to buy it because of stock out.

S. Sharma, *Inventory Parameters*, DOI 10.1007/978-981-10-4545-5_7

(b) Nature/behavior of the customer who could not get the product, i.e., whether this person will tell many potential customers regarding nonavailability and there might be potential loss of other customers also.
(c) Does it affect the brand name concerning that product as well as in case of a large retailer?
(d) Skill of the sales/marketing personnel also affects the cost of stock out. Depending on the skills/interaction with the customer, the sales person can convince the buyer to visit the retailer after some time so that the availability of the product is ensured. In such a situation, there is an effort to make the stock out cost lower.
(e) For how much time, shortages for a product exist? Shortages might be for the duration in:

    (i) Hours
    (ii) Days
    (iii) Weeks

Stock out cost relates to the time. Stock out cost may be estimated per period for one unit of product.

The above-mentioned factors are also represented by Fig. 7.1.

An estimation of the stock out/shortage cost is used along with other input parameters for furthermore analysis. Such case is shown in Fig. 7.2 where stock out/shortages take place during the time $t_2 - t_3$.

Example 7.1
Consider the following values for various input parameters including the annual shortage cost per unit (*K*):

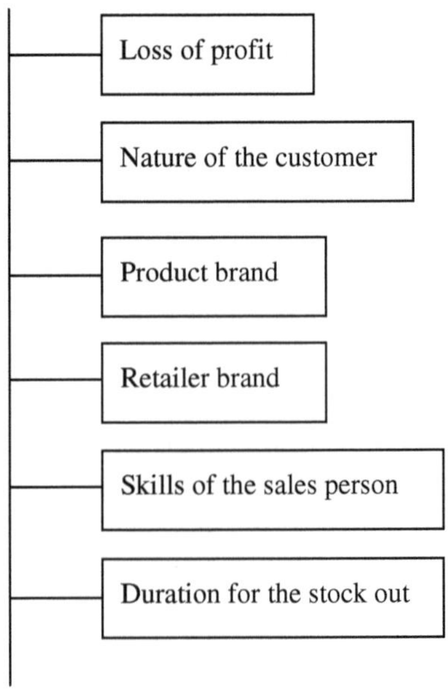

Fig. 7.1 Some aspects pertaining to the cost of stock out

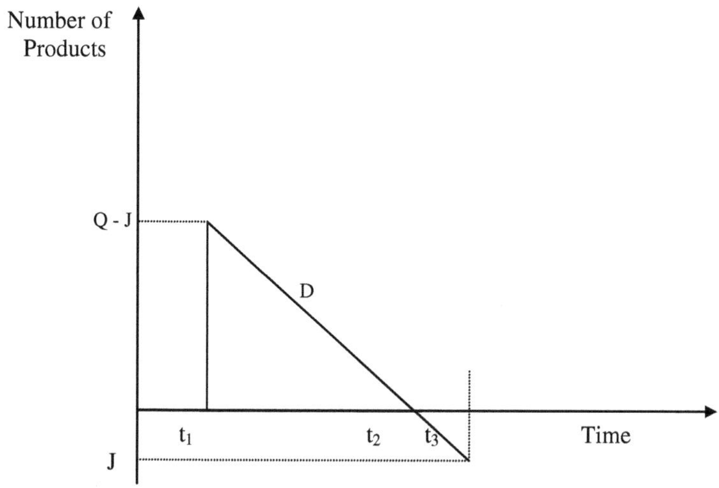

Fig. 7.2 Stock out/shortage case

D	*C*	*I*	*K*
600	30	10	200

When the shortages/backorders are included, the output parameters are lot size, stock out/shortage quantity, and related cost. From Eqs. (1.1), (1.2), and (1.3), the output parameters are:

$$Q^* = \sqrt{\frac{2DC(K+I)}{IK}}$$

$$J^* = \sqrt{\frac{2DCI}{K(K+I)}}$$

$$E^* = \sqrt{\frac{2DCIK}{(K+I)}}$$

With the use of above equations, the following values for output parameters are obtained:

$$Q = 61.4817$$
$$J = 2.93$$
$$E = ₹\ 585.54$$

In order to study a variation in the shortage cost, it is increased/decreased by certain percentage.

7.2 Shortage Cost Increase

Shortage cost may be increased because of the following reasons:

(i) When the retailer earns certain percentage profit on the price of a product and price increases owing to certain economic aspect, then the profit per unit product increases. Because of profit increase and nonavailability of product, loss of profit is at a relatively higher level and stock out cost becomes higher.
(ii) In case of formal or informal consumer groups, dissatisfaction of one consumer may get extended to whole group directly or indirectly at least partially. This may result into higher stock out or shortage cost.
(iii) Sales person is unable to satisfy the consumer on account of nonavailability of product or unable to successfully promise the availability at a later date. In such case also, there is a chance of increase in shortage cost.
(iv) In case of a seasonal product, shortages may lead to the loss of comparatively higher consumer demands, thus making the shortage cost higher.
(v) If shortage period is prolonged, then there is a possibility of higher stock out cost.

These reasons are also summarized in Fig. 7.3.

In the context of optimizing the scenario with an increase in shortage cost, the following effects may be observed:

(a) Decrease in the lot size
(b) Decrease in the maximum stock out quantity
(c) Increase in the total related cost

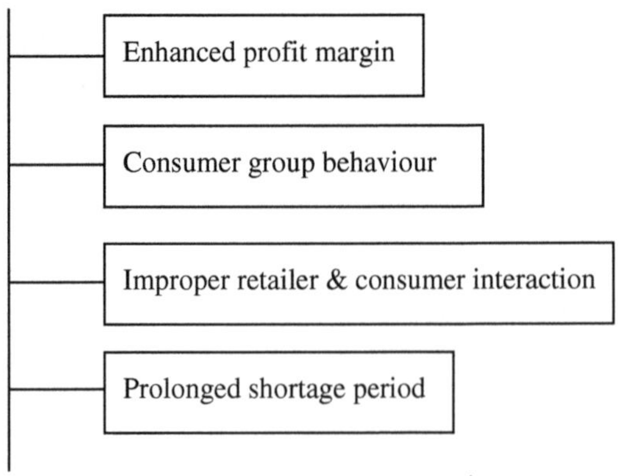

Fig. 7.3 Factors associated with the shortage cost increase

Example 7.2
From the Example 7.1, consider the following data:

D	C	I	K	Q	E	J
600	30	10	200	61.4817	585.54	2.93

Implement an increase in annual shortage cost per unit (K) as follows:

% Increase in K	5%	10%	15%	20%	25%	30%
K	210	220	230	240	250	260

The effects are as shown in Table 7.1.
For a generalization, let:

P = % increase in stock out cost, and therefore,

$$K_1 = K\left(1 + \frac{P}{100}\right)$$

A decrease in the lot size with reference to an increased K_1:

$$= \sqrt{\frac{2DC(K+I)}{IK}} - \sqrt{\frac{2DC(K_1+I)}{IK_1}}$$

$$= \sqrt{\frac{2DC(K+I)}{IK}}\left[1 - \sqrt{\frac{K(K_1+I)}{K_1(K+I)}}\right]$$

$$= \sqrt{\frac{2DC(K+I)}{IK}}\left[1 - \sqrt{\frac{K(1+P/100)+I}{(1+P/100)(K+I)}}\right]$$

And:

$$\%\,\text{decrease in lot size} = 1 - \sqrt{\frac{K(1+P/100)+I}{(1+P/100)(K+I)}}$$

Table 7.1 Effects on parameters corresponding to stock out cost increase

% Increase in K	5%	10%	15%	20%	25%	30%
K	210	220	230	240	250	260
Q	61.4	61.35	61.29	61.24	61.19	61.14
% Decrease in Q	0.1%	0.2%	0.3%	0.4%	0.5%	0.6%
J	2.79	2.67	2.55	2.45	2.35	2.26
% Decrease in J	4.8%	8.9%	13.0%	16.4%	19.8%	22.9%
E	586.2	586.81	587.37	587.88	588.35	588.78
% Increase in E	0.1%	0.2%	0.3%	0.4%	0.5%	0.6%

$$\text{Decrease in the optimum stock out units} = \sqrt{\frac{2DCI}{K(K+I)}} - \sqrt{\frac{2DCI}{K_1(K_1+I)}}$$

$$= \sqrt{\frac{2DCI}{K(K+I)}}\left[1 - \sqrt{\frac{K(K+I)}{K_1(K_1+I)}}\right]$$

$$= \sqrt{\frac{2DCI}{K(K+I)}}\left[1 - \sqrt{\frac{(K+I)}{(1+P/100)\{K(1+P/100)+I\}}}\right]$$

And

$$\%\ \text{decrease in optimum stock out quantity} = 1 - \sqrt{\frac{(K+I)}{(1+P/100)\{K(1+P/100)+I\}}}$$

Increase in the total related cost:

$$= \sqrt{\frac{2DCIK_1}{(K_1+I)}} - \sqrt{\frac{2DCIK}{(K+I)}}$$

$$= \sqrt{\frac{2DCIK}{(K+I)}}\left[\sqrt{\frac{K_1(K+I)}{K(K_1+I)}} - 1\right]$$

$$= \sqrt{\frac{2DCIK}{(K+I)}}\left[\sqrt{\frac{(1+P/100)(K+I)}{K(1+P/100)+I}} - 1\right]$$

And:

$$\%\ \text{increase in cost} = \sqrt{\frac{(1+P/100)(K+I)}{K(1+P/100)+I}} - 1$$

The derived results are also summarized in Table 7.2.

% decrease in the stock out quantity is higher than that in the lot size. This is because:

$$1 - \sqrt{\frac{(K+I)}{(1+P/100)\{K(1+P/100)+I\}}} > 1 - \sqrt{\frac{K(1+P/100)+I}{(1+P/100)(K+I)}}$$

Table 7.2 Results with reference to an increased stock out cost

Decrease in the lot size	$\sqrt{\frac{2DC(K+I)}{IK}}\left[1-\sqrt{\frac{K(1+P/100)+I}{(1+P/100)(K+I)}}\right]$
% Decrease in the lot size	$1-\sqrt{\frac{K(1+P/100)+I}{(1+P/100)(K+I)}}$
Additional related cost	$\sqrt{\frac{2DCIK}{(K+I)}}\left[\sqrt{\frac{(1+P/100)(K+I)}{K(1+P/100)+I}}-1\right]$
% Increase in cost	$\sqrt{\frac{(1+P/100)(K+I)}{K(1+P/100)+I}}-1$
Decrease in the stock out units	$\sqrt{\frac{2DCI}{K(K+I)}}\left[1-\sqrt{\frac{(K+I)}{(1+P/100)\{K(1+P/100)+I\}}}\right]$
% Decrease in the stock out units	$1-\sqrt{\frac{(K+I)}{(1+P/100)\{K(1+P/100)+I\}}}$

Or

$$\sqrt{\frac{K(1+P/100)+I}{(1+P/100)(K+I)}} > \sqrt{\frac{(K+I)}{(1+P/100)\{K(1+P/100)+I\}}}$$

Or

$$\sqrt{\frac{K(1+P/100)+I}{(K+I)}} > \sqrt{\frac{(K+I)}{\{K(1+P/100)+I\}}}$$

Or

$$K(1 + P/100)+I > K+I$$

Or

$$K(1 + P/100) > K$$

and that is true.

7.3 Shortage Cost Decrease

Following are the possible reasons for reduction in the shortage cost:

(i) The profit per unit product may be broadly associated with the difference in selling price and purchase cost of one product. Because of some competition at the market place, selling price might be lower and therefore profit margin may be lower. In some cases, purchase or procurement cost might be higher because of an increase in transportation cost among other factors. With the similar

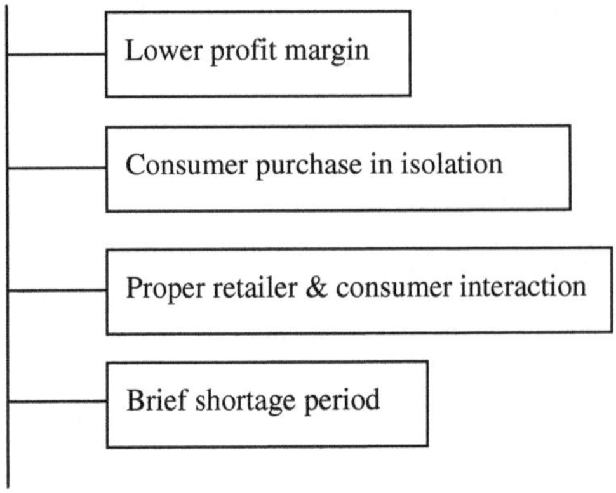

Fig. 7.4 Factors pertaining to the reduced shortage cost

selling price, profit margin may be lower. In case of lower selling price in addition to higher purchase cost, profit margin might be lower furthermore. Under such situations, loss of profit is reduced at least for such product in the event of stock out or shortage, and therefore, the shortage cost decreases.

(ii) Suppose that most of the consumers buy in isolation. On a particular day, a consumer faces shortage, and the next day, retailer is able to fulfill the shortage, i.e., the product is available. In such case, the next day sales might not be affected that much because the previous day consumer may not share the experience with other potential consumers. In such a scenario, shortage cost may be comparatively lower.

(iii) A consumer faces shortage on a particular day and the retailer is able to invite the consumer next day when product would be made available. This way, the effects of shortages in terms of cost might be considerably lower resulting in a lower stock out cost.

(iv) If the duration in which stock out takes place is compressed, then there is a chance of decrease in the effective shortage cost.

These factors are also shown in Fig. 7.4.

Next example is discussed for an illustration of reduction in annual shortage cost per unit (K).

Example 7.3

Consider the following data:

D	C	I	K	Q	E	J
600	30	10	200	61.4817	585.54	2.9277

Implement a decrease in K as follows:

% Decrease in K	5%	10%	15%	20%	25%	30%
K	190	180	170	160	150	140

Table 7.3 Effects on parameters corresponding to stock out cost decrease

% Decrease in K	5%	10%	15%	20%	25%	30%
K	190	180	170	160	150	140
Q	61.56	61.64	61.74	61.85	61.97	62.11
% Increase in Q	0.13%	0.26%	0.42%	0.59%	0.79%	1.02%
J	3.08	3.24	3.43	3.64	3.87	4.14
% Increase in J	5.13%	10.82%	17.16%	24.26%	32.29%	41.42%
E	584.81	584.00	583.10	582.09	580.95	579.66
% Decrease in E	0.13%	0.26%	0.42%	0.59%	0.78%	1.01%

The results are as shown in Table 7.3. The effects on output parameters are as follows:

(i) Increase in lot size
(ii) Increase in shortage or stock out quantity
(iii) Decrease in total related cost

For a generalization, let:

P = % reduction in stock out cost, and therefore,

$$K_1 = K\left(1 - \frac{P}{100}\right)$$

An increase in the lot size with reference to a reduced K_1:

$$= \sqrt{\frac{2DC(K_1 + I)}{IK_1}} - \sqrt{\frac{2DC(K + I)}{IK}}$$

$$= \sqrt{\frac{2DC(K + I)}{IK}}\left[\sqrt{\frac{K(K_1 + I)}{K_1(K + I)}} - 1\right]$$

$$= \sqrt{\frac{2DC(K + I)}{IK}}\left[\sqrt{\frac{K(1 - P/100) + I}{(1 - P/100)(K + I)}} - 1\right]$$

And:

$$\%\,\text{increase in lot size} = \sqrt{\frac{K(1 - P/100) + I}{(1 - P/100)(K + I)}} - 1$$

$$\text{Increase in the optimum stock out units} = \sqrt{\frac{2DCI}{K_1(K_1+I)}} - \sqrt{\frac{2DCI}{K(K+I)}}$$
$$= \sqrt{\frac{2DCI}{K(K+I)}}\left[\sqrt{\frac{K(K+I)}{K_1(K_1+I)}} - 1\right]$$
$$= \sqrt{\frac{2DCI}{K(K+I)}}\left[\sqrt{\frac{(K+I)}{(1-P/100)\{K(1-P/100)+I\}}} - 1\right]$$

And:

$$\%\,\text{increase in optimum stock out quantity} = \sqrt{\frac{(K+I)}{(1-P/100)\{K(1-P/100)+I\}}} - 1$$

Decrease in the total related cost:

$$= \sqrt{\frac{2DCIK}{(K+I)}} - \sqrt{\frac{2DCIK_1}{(K_1+I)}}$$
$$= \sqrt{\frac{2DCIK}{(K+I)}}\left[1 - \sqrt{\frac{K_1(K+I)}{K(K_1+I)}}\right]$$
$$= \sqrt{\frac{2DCIK}{(K+I)}}\left[1 - \sqrt{\frac{(1-P/100)(K+I)}{K(1-P/100)+I}}\right]$$

And:

$$\%\,\text{decrease in cost} = 1 - \sqrt{\frac{(1-P/100)(K+I)}{K(1-P/100)+I}}$$

The derived results are also summarized in Table 7.4.

Table 7.4 Results with reference to a decreased stock out cost

Increase in the lot size	$\sqrt{\frac{2DC(K+I)}{IK}}\left[\sqrt{\frac{K(1-P/100)+I}{(1-P/100)(K+I)}} - 1\right]$
% Increase in the lot size	$\sqrt{\frac{K(1-P/100)+I}{(1-P/100)(K+I)}} - 1$
Decrease in the related cost	$\sqrt{\frac{2DCIK}{(K+I)}}\left[1 - \sqrt{\frac{(1-P/100)(K+I)}{K(1-P/100)+I}}\right]$
% Decrease in cost	$1 - \sqrt{\frac{(1-P/100)(K+I)}{K(1-P/100)+I}}$
Increase in the stock out units	$\sqrt{\frac{2DCI}{K(K+I)}}\left[\sqrt{\frac{(K+I)}{(1-P/100)\{K(1-P/100)+I\}}} - 1\right]$
% Increase in the stock out units	$\sqrt{\frac{(K+I)}{(1-P/100)\{K(1-P/100)+I\}}} - 1$

% increase in lot size is lower than that in the stock out quantities. This is because:

$$\sqrt{\frac{K(1-P/100)+I}{(1-P/100)(K+I)}}-1<\sqrt{\frac{(K+I)}{(1-P/100)\{K(1-P/100)+I\}}}-1$$

Or

$$\sqrt{\frac{K(1-P/100)+I}{(1-P/100)(K+I)}}<\sqrt{\frac{(K+I)}{(1-P/100)\{K(1-P/100)+I\}}}$$

Or

$$\sqrt{\frac{K(1-P/100)+I}{(K+I)}}<\sqrt{\frac{(K+I)}{\{K(1-P/100)+I\}}}$$

Or

$$K(1-P/100)+I<(K+I)$$

Or

$$K(1-P/100)<K$$

And that is true.

7.4 Interaction with the Parameters

With an increase in shortage cost, procurement lot size gets reduced along with an increased total cost. Depending on the objective, i.e., having similar lot size or similar total cost, the interaction with the following parameters may be analyzed:

(a) Demand
(b) Ordering cost
(c) Holding cost

Similarly, with a decrease in shortage cost also, relevant parameters might be analyzed concerning a suitable objective.

7.4.1 *Interaction with the Demand*

With the following information:

D	*C*	*I*	*K*	*Q*	*E*	*J*
600	30	10	200	61.48	585.54	2.93

Consider that the annual shortage cost per unit (*K*) is increased to ₹380. A reduced procurement lot size can be obtained as:

$$Q = \sqrt{\frac{2DC(K+I)}{IK}}$$
$$= \sqrt{\frac{2 \times 600 \times 30(380+10)}{10 \times 380}}$$
$$= 60.78 \text{ units}$$

If the objective is to have similar lot size, a potentially increased demand can be found as:

$$\sqrt{\frac{2 \times D_1 \times 30 \times (380+10)}{10 \times 380}} = 61.48$$

Or

$$D_1 = 613.81$$

For a general approach, let:

P = % increase in stock out cost
M = % increase in demand
Therefore:

$$K_1 = \left(1 + \frac{P}{100}\right)K$$

$$D_1 = \left(1 + \frac{M}{100}\right)D$$

For similar lot size:

$$\sqrt{\frac{2D_1C(K_1+I)}{IK_1}} = \sqrt{\frac{2DC(K+I)}{IK}}$$

Or

$$\frac{2D_1C(K_1+I)}{IK_1}=\frac{2DC(K+I)}{IK}$$

Or

$$\frac{2D(1+M/100)C[K(1+P/100)+I]}{I(1+P/100)K}=\frac{2DC(K+I)}{IK}$$

Or

$$\frac{(1+M/100)[K(1+P/100)+I]}{(1+P/100)}=(K+I)$$

Or

$$\left(1+\frac{M}{100}\right)=\frac{(K+I)(1+P/100)}{K(1+P/100)+I}$$

Or

$$\frac{M}{100}=\frac{(K+I)(1+P/100)}{K(1+P/100)+I}-1$$

Or

$$\frac{M}{100}=\frac{K(1+P/100)+I(1+P/100)-K(1+P/100)-I}{K(1+P/100)+I}$$

Or

$$\frac{M}{100}=\frac{I(1+P/100)-I}{K(1+P/100)+I}$$

Or

$$\frac{M}{100}=\frac{(IP/100)}{K(1+P/100)+I}$$

Or

$$M=\frac{IP}{K(1+P/100)+I}$$

For the current situation, variation of M is shown in Table 7.5 with respect to P. The values are very much lower than the corresponding values of P.

In case of the stock out cost reduction, demand increase might be a management response for similar total cost level.

Consider that the annual shortage cost per unit (K) is reduced to ₹120.

Table 7.5 Variation of M (related to D) with respect to P for similar lot size

S. No.	P	$M = \frac{IP}{K(1+P/100)+I}$
1	10	0.43
2	20	0.8
3	30	1.11
4	40	1.38
5	50	1.61

If the objective is to have similar cost level, then the increased demand can be found as:

$$\sqrt{\frac{2 \times D_1 \times 30 \times 10 \times 120}{(120+10)}} = 585.54$$

Or

$$D_1 = 619$$

In order to generalize, the notation is as follows:

P = % reduction in stock out cost.
M = % increase in the demand.

Therefore:

$$K_1 = \left(1 - \frac{P}{100}\right)K$$

$$D_1 = \left(1 + \frac{M}{100}\right)D$$

For similar total cost:

$$\sqrt{\frac{2D_1CIK_1}{(K_1+I)}} = \sqrt{\frac{2DCIK}{(K+I)}}$$

Or

$$\frac{D_1K_1}{(K_1+I)} = \frac{DK}{(K+I)}$$

Or

$$D_1 = \frac{DK(K_1+I)}{K_1(K+I)}$$

Table 7.6 Variation of M (related to D) with respect to P for similar total cost

S. No.	P	$M = \frac{IP}{(K+I)(1-P/100)}$
1	10	0.53
2	20	1.19
3	30	2.04
4	40	3.17
5	50	4.76

Or

$$1 + \frac{M}{100} = \frac{K(1 - P/100) + I}{(K + I)(1 - P/100)}$$

Or

$$\frac{M}{100} = \frac{K(1 - P/100) + I - (K + I)(1 - P/100)}{(K + I)(1 - P/100)}$$

Or

$$\frac{M}{100} = \frac{I - I(1 - P/100)}{(K + I)(1 - P/100)}$$

Or

$$M = \frac{IP}{(K + I)(1 - P/100)}$$

For the current situation, variation of M is shown in Table 7.6 with respect to P. The values of M are more sensitive toward higher values of P.

7.4.2 Interaction with the Ordering Cost

Consider the data as follows:

D	C	I	K	Q	E	J
600	30	10	200	61.48	585.54	2.93

With an increased stock out cost = ₹380, the total cost is obtained as follows:

$$E = \sqrt{\frac{2DCIK}{+I)}}$$

$$= \sqrt{\frac{2 \times 600 \times 30 \times 10 \times 380}{(380 + 10)}}$$

$$= ₹592.26$$

If the objective is to have similar cost level, then the reduced ordering cost can be found as follows:

$$\sqrt{\frac{2 \times 600 \times C_1 \times 10 \times 380}{(380 + 10)}} = 585.54$$

Or

$$C_1 = ₹29.32$$

In order to generalize, the additional notation is as follows:

P = % increase in stock out cost
M = % reduction in the ordering cost

Therefore:

$$K_1 = \left(1 + \frac{P}{100}\right)K$$

$$C_1 = \left(1 - \frac{M}{100}\right)C$$

For similar total cost:

$$\sqrt{\frac{2DC_1IK_1}{(K_1 + I)}} = \sqrt{\frac{2DCIK}{(K + I)}}$$

Or

$$\frac{C_1K_1}{(K_1 + I)} = \frac{CK}{(K + I)}$$

Or

$$C_1 = \frac{CK(K_1 + I)}{K_1(K + I)}$$

Or

$$\left(1 - \frac{M}{100}\right) = \frac{K(K_1 + I)}{K_1(K + I)}$$

Or

$$1 - \frac{M}{100} = \frac{K(1 + P/100) + I}{(1 + P/100)(K + I)}$$

Or

$$\frac{M}{100} = 1 - \frac{K(1 + P/100) + I}{(1 + P/100)(K + I)}$$

Or

$$\frac{M}{100} = \frac{(1 + P/100)(K + I) - K(1 + P/100) - I}{(1 + P/100)(K + I)}$$

Or

$$\frac{M}{100} = \frac{(1 + P/100)I - I}{(1 + P/100)(K + I)}$$

Or

$$\frac{M}{100} = \frac{(IP/100)}{(1 + P/100)(K + I)}$$

Or

$$M = \frac{IP}{(1 + P/100)(K + I)}$$

For the current situation, variation of M is shown in Table 7.7 with respect to P.

In case of the stock out cost decrease, a reduction in the ordering cost might be an option for having similar lot size.

Consider that the annual shortage cost per unit (K) is reduced to ₹120.

A procurement lot size can be obtained as follows:

$$Q = \sqrt{\frac{2DC(K + I)}{IK}}$$
$$= \sqrt{\frac{2 \times 600 \times 30(120 + 10)}{10 \times 120}}$$
$$= 62.45 \text{ units}$$

If the objective is to have similar lot size, a reduced ordering cost can be found as follows:

Table 7.7 Variation of M (related to C) with respect to P for similar total cost

S. No.	P	$M = \frac{IP}{(1+P/100)(K+I)}$
1	10	0.43
2	20	0.79
3	30	1.099
4	40	1.36
5	50	1.59

$$\sqrt{\frac{2 \times 600 \times C_1 \times (120+10)}{10 \times 120}} = 61.48$$

Or

$$C_1 = ₹29.07$$

For a general approach, let:

P = % reduction in stock out cost
M = % reduction in the ordering cost

Therefore:

$$K_1 = \left(1 - \frac{P}{100}\right)K$$

$$C_1 = \left(1 - \frac{M}{100}\right)C$$

For similar lot size:

$$\sqrt{\frac{2DC_1(K_1+I)}{IK_1}} = \sqrt{\frac{2DC(K+I)}{IK}}$$

Or

$$C_1 = \frac{CK_1(K+I)}{K(K_1+I)}$$

Or

$$1 - \frac{M}{100} = \frac{(1-P/100)(K+I)}{K(1-P/100)+I}$$

Or

$$\frac{M}{100} = \frac{K(1-P/100)+I-(1-P/100)(K+I)}{K(1-P/100)+I}$$

Or

$$\frac{M}{100} = \frac{I-(1-P/100)I}{K(1-P/100)+I}$$

Table 7.8 Variation of M (related to C) with respect to P for similar lot size

S. No.	P	$M = \frac{IP}{K(1-P/100)+I}$
1	10	0.53
2	20	1.18
3	30	2.00
4	40	3.08
5	50	4.55

Or

$$M = \frac{IP}{K(1 - P/100) + I}$$

For the current situation, variation of M is shown in Table 7.8 with respect to P. The values are more sensitive toward higher values of P.

7.4.3 Interaction with the Holding Cost

With the input and output data:

D	C	I	K	Q	E	J
600	30	10	200	61.48	585.54	2.93

If the value of K is increased from ₹200 to ₹380, the total cost is increased. In case where an objective is to have similar total cost as before, an option might be to decrease the holding cost. In order to find the value of revised holding cost:

$$\sqrt{\frac{2 \times 600 \times 30 \times I_1 \times 380}{(380 + I_1)}} = 585.54$$

Or

$$\frac{2 \times 600 \times 30 \times I_1 \times 380}{(380 + I_1)} = 585.54 \times 585.54$$

Or

$$13680000 I_1 = 342857.09(380 + I_1)$$

Or

$$13680000I_1 - 342857.09I_1 = 342857.09 \times 380$$

Or

$$I_1 = \frac{342857.09 \times 380}{13337143}$$

Or

$$I_1 = ₹9.77$$

For a generalization:

P = % increase in stock out cost
M = % reduction in the holding cost

Therefore:

$$K_1 = \left(1 + \frac{P}{100}\right)K$$

$$I_1 = \left(1 - \frac{M}{100}\right)I$$

For similar total cost:

$$\sqrt{\frac{2DCI_1K_1}{(K_1 + I_1)}} = \sqrt{\frac{2DCIK}{(K + I)}}$$

Or

$$I_1K_1(K + I) = IK(K_1 + I_1) \tag{7.1}$$

Or

$$(1 - M/100)IK(1 + P/100)(K + I) = IK[K(1 + P/100) + I(1 - M/100)]$$

Or

$$(1 - M/100)(1 + P/100)(K + I) = K(1 + P/100) + I(1 - M/100)$$

Or

$$(1 - M/100)[(1 + P/100)(K + I) - I] = K(1 + P/100)$$

Or

$$1 - \frac{M}{100} = \frac{K(1 + P/100)}{(1 + P/100)(K + I) - I}$$

Or

$$1 - \frac{M}{100} = \frac{K(1 + P/100)}{(1 + P/100)K + (1 + P/100)I - I}$$

Or

$$1 - \frac{M}{100} = \frac{K(1 + P/100)}{(1 + P/100)K + (IP/100)}$$

Or

$$\frac{M}{100} = 1 - \frac{K(1 + P/100)}{(1 + P/100)K + (IP/100)}$$

Or

$$\frac{M}{100} = \frac{(1 + P/100)K + (IP/100) - K(1 + P/100)}{(1 + P/100)K + (IP/100)}$$

Or

$$\frac{M}{100} = \frac{(IP/100)}{(1 + P/100)K + (IP/100)}$$

Or

$$M = \frac{IP}{(1 + P/100)K + (IP/100)} \tag{7.2}$$

For the current situation, variation of M is shown in Table 7.9 with respect to P.

If the objective is to have similar lot size, a revised holding cost can be found as follows:

Table 7.9 Variation of M (related to I) with respect to P for similar total cost

S. No.	P	$M = \frac{IP}{(1+P/100)K+(IP/100)}$
1	10	0.45
2	20	0.83
3	30	1.14
4	40	1.41
5	50	1.64

$$\sqrt{\frac{2\times 600\times 30\times (380+I_1)}{I_1\times 380}} = 61.48$$

Or

$$\frac{2\times 600\times 30\times (380+I_1)}{I_1\times 380} = 61.48\times 61.48$$

Or

$$36000(380+I_1) = 1436320.35I_1$$

Or

$$1400320.35I_1 = 36000\times 380$$

Or

$$I_1 = ₹9.77$$

The value of I_1 is equivalent to ₹9.77 and that was also obtained for similar total cost. This is because:

$$\sqrt{\frac{2DC(K_1+I_1)}{I_1K_1}} = \sqrt{\frac{2DC(K+I)}{IK}}$$

Or

$$I_1K_1(K+I) = IK(K_1+I_1)$$

It is similar to the expression (7.1), and therefore, value of M will also be similar to the expression (7.2).

In above cases, a change is triggered by the increase in shortage cost (K). However, this may also be triggered by a variation in other parameters.

7.5 Variation of *K* Triggered by Change in Other Parameters

Because of a change in relevant parameters other than the shortage cost (K), suitable management response can be analyzed by way of quantification of the reduction in K in order to achieve the desired objective, i.e., similar lot size/total cost.

7.5.1 *Demand Variation*

In case of demand increase, the notation is as follows:

P = % increase in demand
M = % reduction in the shortage cost

Therefore:

$$D_1 = \left(1 + \frac{P}{100}\right)D$$

$$K_1 = \left(1 - \frac{M}{100}\right)K$$

For similar total cost:

$$\sqrt{\frac{2D_1CIK_1}{(K_1+I)}} = \sqrt{\frac{2DCIK}{(K+I)}}$$

Or

$$D_1K_1(K+I) = DK(K_1+I)$$

Or

$$(1+P/100)K_1(K+I) = K(K_1+I)$$

Or

$$K_1[(1+P/100)(K+I) - K] = KI$$

Or

$$K_1 = \frac{KI}{I+(P/100)(K+I)}$$

Or

$$1-\frac{M}{100}=\frac{I}{I+(P/100)(K+I)}$$

Or

$$\frac{M}{100}=1-\frac{I}{I+(P/100)(K+I)}$$

Or

$$\frac{M}{100}=\frac{(P/100)(K+I)}{I+(P/100)(K+I)}$$

Or

$$M=\frac{P(K+I)}{I+(P/100)(K+I)} \tag{7.3}$$

With an unavoidable decrease in demand:

P = % decrease in demand
M = % reduction in the shortage cost

Therefore:

$$D_1=D\left(1-\frac{P}{100}\right)$$

$$K_1=K\left(1-\frac{M}{100}\right)$$

For similar lot size:

$$\sqrt{\frac{2D_1C(K_1+I)}{IK_1}}=\sqrt{\frac{2DC(K+I)}{IK}}$$

Or

$$KD_1(K_1+I)=DK_1(K+I)$$

Or

$$K(1 - P/100)(K_1 + I) = K_1(K + I)$$

Or

$$K_1[(K + I) - K(1 - P/100)] = IK(1 - P/100)$$

Or

$$K_1 = \frac{IK(1 - P/100)}{I + (KP/100)}$$

Or

$$1 - \frac{M}{100} = \frac{I(1 - P/100)}{I + (KP/100)}$$

Or

$$\frac{M}{100} = 1 - \frac{I(1 - P/100)}{I + (KP/100)}$$

Or

$$\frac{M}{100} = \frac{(K + I)(P/100)}{I + (KP/100)}$$

Or

$$M = \frac{P(K + I)}{I + (KP/100)} \tag{7.4}$$

While comparing with the expression (7.3), denominator of the R.H.S. is lower in this case. Although objective is different, however, the values of M in this case would be comparatively higher.

7.5.2 Ordering Cost Variation

With ordering cost increase, the following notation is considered:

P = % increase in ordering cost
M = % reduction in the shortage cost

Therefore:

$$C_1 = C\left(1 + \frac{P}{100}\right)$$

$$K_1 = K\left(1 - \frac{M}{100}\right)$$

For similar total cost:

$$\sqrt{\frac{2DC_1IK_1}{(K_1 + I)}} = \sqrt{\frac{2DCIK}{(K + I)}}$$

On solving, it can be shown that the similar result as expressed by (7.3) is obtained.

Now, with decrease in ordering cost :

P = % decrease in ordering cost
M = % reduction in the shortage cost

Therefore:

$$C_1 = C\left(1 - \frac{P}{100}\right)$$

$$K_1 = K\left(1 - \frac{M}{100}\right)$$

For similar lot size:

$$\sqrt{\frac{2DC_1(K_1 + I)}{IK_1}} = \sqrt{\frac{2DC(K + I)}{IK}}$$

On solving, it can be shown that the similar result as expressed by (7.4) is obtained.

7.5.3 Holding Cost Increase

In case of the relevant variation in holding cost, i.e., an increase in the parameter, consider the following notation:

P = % increase in holding cost
M = % reduction in the shortage cost

Therefore:

$$I_1 = I\left(1 + \frac{P}{100}\right)$$

$$K_1 = K\left(1 - \frac{M}{100}\right)$$

For similar lot size:

$$\sqrt{\frac{2DC(K_1 + I_1)}{I_1 K_1}} = \sqrt{\frac{2DC(K + I)}{IK}}$$

Or

$$I_1 K_1 (K + I) = IK(K_1 + I_1) \tag{7.5}$$

Or

$$K_1[I_1(K + I) - IK] = IKI_1$$

Or

$$K_1 = \frac{IKI_1}{I_1(K + I) - IK}$$

Or

$$1 - \frac{M}{100} = \frac{I(1 + P/100)}{(1 + P/100)(K + I) - K}$$

Or

$$\frac{M}{100} = \frac{(1 + P/100)(K + I) - K - I(1 + P/100)}{(1 + P/100)(K + I) - K}$$

Or

$$\frac{M}{100} = \frac{(1 + P/100)K - K}{I + (P/100)(K + I)}$$

Or

$$\frac{M}{100} = \frac{(KP/100)}{I + (P/100)(K + I)}$$

Or

$$M = \frac{KP}{I + (P/100)(K + I)} \tag{7.6}$$

For similar total cost:

$$\sqrt{\frac{2DCI_1K_1}{(K_1 + I_1)}} = \sqrt{\frac{2DCIK}{(K + I)}}$$

Or

$$I_1K_1(K + I) = IK(K_1 + I_1)$$

As this is similar to the expression (7.5), the derived value of M is similar to the expression (7.6).

7.6 Response with a Combination of Parameters

When there is a possibility of response of the management concerning multiple parameters, then it needs to be analyzed corresponding to variation triggered by a specific parameter. Analysis also corresponds to the aim such as follows:

(i) Similar lot size
(ii) Similar total cost

7.6.1 *With an Aim of Similar Lot Size*

When change is triggered by a specific parameter, lot size varies. With an objective of similar lot size, the following situations are analyzed with initiation from:

(a) Ordering cost
(b) Carrying cost
(c) Stock out cost
(d) Demand

7.6.1.1 Change Triggered by Ordering Cost

When ordering cost is reduced, a response is considered related to a combination of any two parameter variation from the following:

(i) Demand increase
(ii) Carrying cost reduction
(iii) Stock out cost reduction

With reference to above, the combinations (that include stock out cost) are formed as follows:

(a) Demand increase and stock out cost reduction
(b) Carrying cost and stock out cost reduction

Now, each case is discussed.

(a) **Demand increase and stock out cost reduction**:

With the following parameters:
Annual demand, D = 600 units
Ordering cost, C = ₹30
Annual inventory carrying cost per unit, I = ₹10
Annual stock out cost per unit, K = ₹200
Optimum procurement lot size is obtained as:

$$\begin{aligned} Q &= \sqrt{\frac{2DC(K+I)}{IK}} \\ &= \sqrt{\frac{2 \times 600 \times 30(200+10)}{10 \times 200}} \\ &= 61.48 \text{ units} \end{aligned}$$

Now, if the change in parameter is triggered by a reduction in ordering cost by 5%, then:

$$C_1 = ₹28.5$$

In order to have similar lot size, a potentially increased demand can be found as follows:

$$\sqrt{\frac{2 \times D_1 \times 28.5 \times (200+10)}{10 \times 200}} = 61.48$$

Or

$$D_1 = 631.54$$

That is 5.26% increase in demand.

However, if it is not possible to increase the demand by more than 4% (say), then a combination of demand and stock out cost might be considered for variation. For this case (considering max. increase in D as 4%):

$$\sqrt{\frac{2 \times 624 \times 28.5 \times (10+K_1)}{10K_1}} = 61.48$$

Or

$$K_1 = 159.50$$

That is approx. 20.3% reduction in the stock out cost.

For a general approach, let:

$$C_1 = C\left(1 - \frac{P}{100}\right)$$

$$D_1 = D\left(1 + \frac{M}{100}\right)$$

$$K_1 = K\left(1 - \frac{N}{100}\right)$$

For similar lot size:

$$\sqrt{\frac{2D_1C_1(K_1 + I)}{IK_1}} = \sqrt{\frac{2DC(K + I)}{IK}}$$

Or

$$K_1DC(K + I) = KD_1C_1(K_1 + I)$$

Or

$$K_1[DC(K + I) - KD_1C_1] = KID_1C_1$$

Or

$$K_1 = \frac{KID_1C_1}{DC(K + I) - KD_1C_1}$$

Or

$$1 - \frac{N}{100} = \frac{ID_1C_1}{DC(K + I) - KD_1C_1}$$

Or

$$1 - \frac{N}{100} = \frac{I(1 + M/100)(1 - P/100)}{K + I - K(1 + M/100)(1 - P/100)}$$

Or

$$\frac{N}{100} = \frac{K + I - K(1 + M/100)(1 - P/100) - I(1 + M/100)(1 - P/100)}{K + I - K(1 + M/100)(1 - P/100)}$$

Or

$$N = \frac{100(K + I)[1 - (1 + M/100)(1 - P/100)]}{I + K[1 - (1 + M/100)(1 - P/100)]}$$

For $P = 5$,

$$N = \frac{100(K + I)[1 - 0.95(1 + M/100)]}{I + K[1 - 0.95(1 + M/100)]}$$

For the Example data:

$$N = \frac{105 - 19.95M}{2 - 0.19M}$$

The values of N (for $P = 5$) are shown in Table 7.10 corresponding to the different values of M. The values of N are more sensitive toward higher values of M.

(b) **Carrying cost and stock out cost reduction**:

$$C_1 = C\left(1 - \frac{P}{100}\right)$$

$$I_1 = I\left(1 - \frac{M}{100}\right)$$

$$K_1 = K\left(1 - \frac{N}{100}\right)$$

For similar lot size:

$$\sqrt{\frac{2DC_1(K_1 + I_1)}{I_1K_1}} = \sqrt{\frac{2DC(K + I)}{IK}}$$

Table 7.10 N (for K) corresponding to M (for D) concerning P (for C) = 5

S. No.	M	$N = \frac{105-19.95M}{2-0.19M}$
1	1	46.99
2	2	40.19
3	3	31.57
4	4	20.32

Or

$$K_1 I_1 C(K + I) = IKC_1(K_1 + I_1)$$

Or

$$K_1[I_1 C(K + I) - C_1 IK] = IKC_1 I_1$$

Or

$$K_1 = \frac{IKC_1 I_1}{I_1 C(K + I) - C_1 IK}$$

Or

$$1 - \frac{N}{100} = \frac{C_1 I_1}{(1 - M/100)C(K + I) - C_1 K}$$

Or

$$1 - \frac{N}{100} = \frac{C(1 - P/100)I(1 - M/100)}{(1 - M/100)C(K + I) - KC(1 - P/100)}$$

Or

$$\frac{N}{100} = 1 - \frac{I(1 - P/100)(1 - M/100)}{(1 - M/100)(K + I) - K(1 - P/100)}$$

Or

$$N = 100\left[1 - \frac{I(1 - P/100)(1 - M/100)}{(1 - M/100)(K + I) - K(1 - P/100)}\right]$$

For $P = 5$:

$$N = \frac{1050 - 200.5M}{20 - 2.1M}$$

The values of N (for $P = 5$) are shown in Table 7.11 corresponding to the different values of M. The values of N are more sensitive toward higher values of M.

Table 7.11 N (for K) corresponding to M (for I) concerning P (for C) = 5

S. No.	M	$N = \frac{1050-200.5M}{20-2.1M}$
1	1	47.46
2	2	41.08
3	3	32.74
4	4	21.38

7.6.1.2 Change Triggered by Carrying Cost

When carrying cost is increased, a response comprises of the following parameter variation:

(i) Demand increase
(ii) Stock out cost reduction

Demand increase and stock out cost reduction:

$$I_1 = I\left(1 + \frac{P}{100}\right)$$

$$D_1 = D\left(1 + \frac{M}{100}\right)$$

$$K_1 = K\left(1 - \frac{N}{100}\right)$$

For similar lot size:

$$\sqrt{\frac{2D_1C(K_1 + I_1)}{I_1K_1}} = \sqrt{\frac{2DC(K+I)}{IK}}$$

Or

$$K_1I_1D(K+I) = IKD_1(K_1 + I_1)$$

Or

$$K_1[I_1D(K+I) - IKD_1] = IKD_1I_1$$

Or

$$K_1 = \frac{IKD_1I_1}{DI_1(K+I) - IKD_1}$$

Or

$$1 - \frac{N}{100} = \frac{ID_1I_1}{DI_1(K+I) - IKD_1}$$

Or

$$1 - \frac{N}{100} = \frac{ID(1 + M/100)I(1 + P/100)}{DI(1 + P/100)(K+I) - IKD(1 + M/100)}$$

Table 7.12 N (for K) corresponding to M (for D) concerning P (for I) = 5

S. No.	M	$N = \frac{1000-210.5M}{20.5-2M}$
1	1	42.68
2	2	35.09
3	3	25.41
4	4	12.64

Or

$$1 - \frac{N}{100} = \frac{I(1+M/100)(1+P/100)}{(1+P/100)(K+I) - K(1+M/100)}$$

Or

$$N = 100\left[1 - \frac{I(1+M/100)(1+P/100)}{(1+P/100)(K+I) - K(1+M/100)}\right]$$

For P = 5:

$$N = \frac{1000 - 210.5M}{20.5 - 2M}$$

The values of N (for P = 5) are shown in Table 7.12 corresponding to the different values of M. The values of N are more sensitive toward higher values of M.

7.6.1.3 Change Triggered by Stock Out Cost

When stock out cost is increased, a response comprises of the following parameter variation:

(i) Demand increase
(ii) Carrying cost reduction

Demand increase and carrying cost reduction:

$$K_1 = K\left(1 + \frac{P}{100}\right)$$

$$D_1 = D\left(1 + \frac{M}{100}\right)$$

$$I_1 = I\left(1 - \frac{N}{100}\right)$$

Now:

$$\sqrt{\frac{2D_1C(K_1+I_1)}{I_1K_1}} = \sqrt{\frac{2DC(K+I)}{IK}}$$

Or

$$I_1K_1D(K+I) = IKD_1(K_1+I_1)$$

Or

$$I_1[K_1D(K+I) - IKD_1] = IKD_1K_1$$

Or

$$I_1 = \frac{IKD_1K_1}{DK_1(K+I) - IKD_1}$$

Or

$$1 - \frac{N}{100} = \frac{KD_1K_1}{DK_1(K+I) - IKD_1}$$

Or

$$1 - \frac{N}{100} = \frac{KD(1+M/100)K(1+P/100)}{DK(1+P/100)(K+I) - IKD(1+M/100)}$$

Or

$$1 - \frac{N}{100} = \frac{K(1+M/100)(1+P/100)}{(1+P/100)(K+I) - I(1+M/100)}$$

Or

$$N = 100\left[1 - \frac{K(1+M/100)(1+P/100)}{(1+P/100)(K+I) - I(1+M/100)}\right]$$

For $P = 90$:

$$N = \frac{900 - 390M}{389 - 0.1M}$$

The results are demonstrated in Table 7.13.

Table 7.13 N (for I) corresponding to M (for D) concerning P (for K) = 90

S. No.	M	$N = \frac{900-390M}{389-0.1M}$
1	0.5	1.81
2	1	1.31
3	1.5	0.81
4	2	0.31

7.6.1.4 Change Triggered by Demand

In case of unavoidable demand decrease, a response consists of the following parameter variation:

(i) Carrying cost reduction
(ii) Stock out cost reduction

Carrying cost and stock out cost reduction:

$$D_1 = D\left(1 - \frac{P}{100}\right)$$

$$I_1 = I\left(1 - \frac{M}{100}\right)$$

$$K_1 = K\left(1 - \frac{N}{100}\right)$$

Now:

$$\sqrt{\frac{2D_1C(K_1 + I_1)}{I_1K_1}} = \sqrt{\frac{2DC(K + I)}{IK}}$$

Or

$$K_1I_1D(K + I) = IKD_1(K_1 + I_1)$$

Or

$$K_1[I_1D(K + I) - IKD_1] = IKD_1I_1$$

Or

$$K_1 = \frac{IKD_1I_1}{DI_1(K + I) - IKD_1}$$

Table 7.14 N (for K) corresponding to M (for I) concerning P (for D) = 10

S. No.	M	$N = \frac{2100-201M}{30-2.1M}$
1	2	65.81
2	4	60.00
3	6	51.38
4	8	37.27
5	10	10.00

Or

$$1 - \frac{N}{100} = \frac{ID_1 I_1}{DI_1(K+I) - IKD_1}$$

Or

$$1 - \frac{N}{100} = \frac{ID(1 - P/100)I(1 - M/100)}{DI(1 - M/100)(K+I) - IKD(1 - P/100)}$$

Or

$$1 - \frac{N}{100} = \frac{I(1 - P/100)(1 - M/100)}{(1 - M/100)(K+I) - K(1 - P/100)}$$

Or

$$N = 100\left[1 - \frac{I(1 - P/100)(1 - M/100)}{(1 - M/100)(K+I) - K(1 - P/100)}\right]$$

For P = 10:

$$N = \frac{2100 - 201M}{30 - 2.1M}$$

The results are demonstrated in Table 7.14. The values of N are more sensitive toward higher values of M.

7.6.2 *With an Aim of Similar Total Cost*

When change is triggered by a specific parameter, total cost varies. With an objective of similar total cost, the following situations are analyzed with initiation from:

(a) Ordering cost
(b) Carrying cost
(c) Stock out cost
(d) Demand

7.6.2.1 Variation Triggered by Ordering Cost

With an increase in the ordering cost, response can be in the form of:

(i) Carrying cost reduction
(ii) Stock out cost reduction

Carrying cost and stock out cost reduction:
With the input and output data:

D	C	I	K	Q	E	J
600	30	10	200	61.48	585.54	2.93

If the value of C is increased from ₹30 to ₹36, the total cost is increased. In case where an objective is to have similar total cost as before, an option might be to decrease the carrying cost. In order to find the value of revised carrying cost:

$$\sqrt{\frac{2 \times 600 \times 36 \times I_1 \times 200}{(200 + I_1)}} = 585.54$$

Or

$$I_1 = 8.26$$

That is a reduction of 17.36%. However, if it is not possible to reduce by more than 15% (say), then a combination of I and K may be explored. Now:

$$\sqrt{\frac{2 \times 600 \times 36 \times 8.5K}{(K + 8.5)}} = 585.54$$

Or

$$K = 119.72$$

That is 40.14% reduction in K.
In order to generalize:

$$C_1 = C\left(1 + \frac{P}{100}\right)$$

$$I_1 = I\left(1 - \frac{M}{100}\right)$$

$$K_1 = K\left(1 - \frac{N}{100}\right)$$

For the similar total cost:

$$\sqrt{\frac{2DC_1I_1K_1}{(K_1 + I_1)}} = \sqrt{\frac{2DCIK}{(K + I)}}$$

Or

$$C_1K_1I_1(K + I) = CIK(K_1 + I_1)$$

Or

$$K_1[C_1I_1(K + I) - CIK] = CIKI_1$$

Or

$$K_1 = \frac{CIKI_1}{C_1I_1(K + I) - CIK}$$

Or

$$1 - \frac{N}{100} = \frac{CII_1}{C_1I_1(K + I) - CIK}$$

Or

$$1 - \frac{N}{100} = \frac{CII(1 - M/100)}{CI(1 + P/100)(1 - M/100)(K + I) - CIK}$$

Or

$$1 - \frac{N}{100} = \frac{I(1 - M/100)}{(1 + P/100)(1 - M/100)(K + I) - K}$$

Or

$$N = 100\left[1 - \frac{I(1 - M/100)}{(1 + P/100)(1 - M/100)(K + I) - K}\right]$$

For $P = 20$:

Table 7.15 N (for K) corresponding to M (for I) concerning P (for C) = 20

S. No.	M	$N = \frac{4200-242M}{52-2.52M}$
1	10	66.42
2	11	63.34
3	12	59.56
4	13	54.78
5	14	48.56
6	15	40.14

$$N = \frac{4200 - 242M}{52 - 2.52M}$$

The results are shown in Table 7.15. The values of N are more sensitive toward higher values of M.

7.6.2.2 Variation Triggered by Carrying Cost

Because of an increase in carrying cost, a response incorporates reduction in (i) ordering cost and (ii) stock out cost.

Ordering cost and stock out cost reduction:

$$I_1 = I\left(1 + \frac{P}{100}\right)$$

$$C_1 = C\left(1 - \frac{M}{100}\right)$$

$$K_1 = K\left(1 - \frac{N}{100}\right)$$

For the similar total cost:

$$\sqrt{\frac{2DC_1I_1K_1}{(K_1 + I_1)}} = \sqrt{\frac{2DCIK}{(K + I)}}$$

Or

$$C_1K_1I_1(K + I) = CIK(K_1 + I_1)$$

Or

$$K_1[C_1I_1(K + I) - CIK] = CIKI_1$$

Or

$$K_1 = \frac{CIKI_1}{C_1I_1(K+I) - CIK}$$

Or

$$1 - \frac{N}{100} = \frac{CII_1}{C_1I_1(K+I) - CIK}$$

Or

$$1 - \frac{N}{100} = \frac{CII(1+P/100)}{CI(1-M/100)(1+P/100)(K+I) - CIK}$$

Or

$$1 - \frac{N}{100} = \frac{I(1+P/100)}{(1-M/100)(1+P/100)(K+I) - K}$$

Or

$$N = 100\left[1 - \frac{I(1+P/100)}{(1-M/100)(1+P/100)(K+I) - K}\right]$$

For $P = 10$:

$$N = \frac{2000 - 231M}{31 - 2.31M}$$

Table 7.16 shows the results. Sensitivity for the values of N increases toward higher values of M.

7.6.2.3 Variation Triggered by Stock Out Cost

Because of an increase in stock out cost, a response incorporates reduction in (i) ordering cost and (ii) carrying cost.

Table 7.16 N (for K) corresponding to M (for C) concerning P (for I) = 10

S. No.	M	$N = \frac{2000-231M}{31-2.31M}$
1	1	61.66
2	2	58.30
3	3	54.30
4	4	49.45
5	5	43.44

Ordering cost and carrying cost reduction:

$$K_1 = K\left(1 + \frac{P}{100}\right)$$

$$C_1 = C\left(1 - \frac{M}{100}\right)$$

$$I_1 = I\left(1 - \frac{N}{100}\right)$$

Now:

$$\sqrt{\frac{2DC_1I_1K_1}{(K_1 + I_1)}} = \sqrt{\frac{2DCIK}{(K + I)}}$$

Or

$$C_1K_1I_1(K + I) = CIK(K_1 + I_1)$$

Or

$$I_1[C_1K_1(K + I) - CIK] = CIKK_1$$

Or

$$I_1 = \frac{CIKK_1}{C_1K_1(K + I) - CIK}$$

Or

$$1 - \frac{N}{100} = \frac{CKK_1}{C_1K_1(K + I) - CIK}$$

Or

$$1 - \frac{N}{100} = \frac{CKK(1 + P/100)}{CK(1 - M/100)(1 + P/100)(K + I) - CIK}$$

Or

$$1 - \frac{N}{100} = \frac{K(1 + P/100)}{(1 - M/100)(1 + P/100)(K + I) - I}$$

Table 7.17 N (for I) corresponding to M (for C) concerning P (for K) = 90

S. No.	M	$N = \frac{900-399M}{389-3.99M}$
1	0.5	1.81
2	1.0	1.30
3	1.5	0.79
4	2.0	0.27

Or

$$N = 100\left[1 - \frac{K(1+P/100)}{(1-M/100)(1+P/100)(K+I) - I}\right]$$

For $P = 90$:

$$N = \frac{900 - 399M}{389 - 3.99M}$$

Table 7.17 shows the results.

7.6.2.4 Variation Triggered by Demand

When demand is increased, a response is considered related to a combination of any two parameter variation from the following:

(i) Ordering cost decrease
(ii) Carrying cost decrease
(iii) Stock out cost decrease

With reference to above, the combinations (that include stock out cost) are formed as follows:

(a) Ordering cost decrease and stock out cost decrease
(b) Carrying cost decrease and stock out cost decrease

Now, each case is discussed.

(a) **Ordering cost decrease and stock out cost decrease**:

$$D_1 = D\left(1 + \frac{P}{100}\right)$$

$$C_1 = C\left(1 - \frac{M}{100}\right)$$

$$K_1 = K\left(1 - \frac{N}{100}\right)$$

For the similar total cost:

$$\sqrt{\frac{2D_1C_1IK_1}{(K_1+I)}} = \sqrt{\frac{2DCIK}{(K+I)}}$$

Or

$$D_1C_1K_1(K+I) = DCK(K_1+I)$$

Or

$$K_1[D_1C_1(K+I) - DCK] = DCKI$$

Or

$$K_1 = \frac{DCKI}{D_1C_1(K+I) - DCK}$$

Or

$$1 - \frac{N}{100} = \frac{DCI}{D_1C_1(K+I) - DCK}$$

Or

$$1 - \frac{N}{100} = \frac{DCI}{DC(1+P/100)(1-M/100)(K+I) - DCK}$$

Or

$$1 - \frac{N}{100} = \frac{I}{(1+P/100)(1-M/100)(K+I) - K}$$

Or

$$N = 100\left[1 - \frac{I}{(1+P/100)(1-M/100)(K+I) - K}\right]$$

For $P = 10$:

$$N = \frac{2100 - 231M}{31 - 2.31M}$$

Table 7.18 shows the results. Sensitivity for the values of N increases toward higher values of M.

Table 7.18 N (for K) corresponding to M (for C) concerning P (for D) = 10

S. No.	M	$N = \frac{2100-231M}{31-2.31M}$
1	1	65.14
2	2	62.09
3	3	58.45
4	4	54.04
5	5	48.59

(b) **Carrying cost decrease and stock out cost decrease**:

$$D_1 = D\left(1 + \frac{P}{100}\right)$$

$$I_1 = I\left(1 - \frac{M}{100}\right)$$

$$K_1 = K\left(1 - \frac{N}{100}\right)$$

Now:

$$\sqrt{\frac{2D_1CI_1K_1}{(K_1 + I_1)}} = \sqrt{\frac{2DCIK}{(K + I)}}$$

Or

$$D_1I_1K_1(K + I) = DIK(K_1 + I_1)$$

Or

$$K_1[D_1I_1(K + I) - DIK] = DIKI_1$$

Or

$$K_1 = \frac{DIKI_1}{D_1I_1(K + I) - DIK}$$

Or

$$1 - \frac{N}{100} = \frac{DII_1}{D_1I_1(K + I) - DIK}$$

Table 7.19 N (for K) corresponding to M (for I) concerning P (for D) = 10

S. No.	M	$N = \frac{2100-221M}{31-2.31M}$
1	1	65.49
2	2	62.85
3	3	59.70
4	4	55.88
5	5	51.16

Or

$$1 - \frac{N}{100} = \frac{DII(1 - M/100)}{DI(1 + P/100)(1 - M/100)(K + I) - DIK}$$

Or

$$1 - \frac{N}{100} = \frac{I(1 - M/100)}{(1 + P/100)(1 - M/100)(K + I) - K}$$

Or

$$N = 100\left[1 - \frac{I(1 - M/100)}{(1 + P/100)(1 - M/100)(K + I) - K}\right]$$

For $P = 10$:

$$N = \frac{2100 - 221M}{31 - 2.31M}$$

The results are shown in Table 7.19. The values of N are more sensitive toward higher values of M.

7.7 Stock Out Cost Estimation for Various Products

In case of the shortages or stock outs, the effects are varying as per the industry/products such as follows:

(i) FMCG
(ii) Pharmaceutical
(iii) Agricultural
(iv) Hi-tech

7.7.1 FMCG

In the business of fast-moving consumer goods, demands are quite considerable. However, sometimes because of certain quality problems, it might be necessary to lift the whole stock from the market. This action might be justified for the sake of brand image of the company. Therefore, it becomes necessary to pay a lot of attention to the associated issues that affect the brand image of the company. Consumers face shortages and goodwill may be lost.

Additionally because of improper supply–demand planning, consumers may face shortages at a particular location and might switch to the competitor's products resulting in profit loss for a relevant FMCG company. In order to compensate for the lost consumers, the company might adopt advertising. However, the advertising expenditure should be taken into account. An analysis of such aspects helps in estimating the stock out costs for FMCG case.

7.7.2 Pharmaceutical

Suppose that a patient or his representative visits the retailer in order to purchase a particular medicine and it is not in stock. Then, there might be the two types of situations:

(i) The customer goes away and contacts another retailer. If the demand is satisfied, then the first retailer lost the profit related to the demand.
(ii) If it is possible and convenient in case of no urgency, the customer is ready to wait for some period, say one week. Retailer ensures the availability of medicine. The customer visits again and purchases the requirement. However, if this happens quite frequently, then the customer may switch the retailer resulting into a kind of permanent loss.

Such situations may arise at the customer–retailer interface. However, at the retailer–distributor/wholesaler or pharmaceutical company interface, if stock out situation is being faced, then the company image might suffer. Another pharmaceutical company might be beneficiary if it is able to fulfill the gap. Stock out costs may affect retailer as well as other companies in the chain. A precise estimation of stock out cost needs to be made after taking into consideration various related aspects.

7.7.3 Agricultural

If agricultural products such as food grains are not available in sufficient quantities at different locations, then the society as a whole suffers a lot. Similarly, vegetables and fruits should also be available. In case of the supply–demand mismatch, i.e., the case when demand exceeds supply, shortages take place and price rise may become an issue. A macrolevel understanding of situation will help in knowing the stock out or shortages issues better.

However, at the microlevel, if retailer is not keeping the enough stock of such items, few consumers may not get it and shortage/stock out scenario happens. Stock out costs need to be estimated in the context of:

(i) Not fulfilling the consumer demands
(ii) Profit loss
(iii) Dissatisfaction of customers
(iv) Long-term loss of some potential consumers

7.7.4 Hi-tech

Because of the shorter product life cycles of certain hi-tech products, it may not be possible to switch supplier in shorter time. In case where a supplier fails to deliver the components/subassemblies on time, it may be difficult for the production company to fabricate such product on time or to achieve their production schedule. This may affect the availability of such product at the dealer level and shortages may take place. Because of the stock out in case of a company, the end consumers may be inclined toward purchase from its competitors in case where a similar product from the competitor company is easily available. While estimating the cost of stock outs to the company, such aspects may be taken into consideration additionally.

Index

A
Agricultural, 40, 42, 68, 108, 109, 232, 233
Annual inventory carrying cost, 11, 12, 14, 15, 17, 27, 32, 49, 71, 73, 84, 86, 91, 95, 97, 102, 105, 111, 133, 134, 143, 145, 215
Annual ordering cost, 14, 15, 17, 133, 143

B
Backlogging situation, 142
Backordering scenario, 80

C
Carrying cost estimation, 71, 72
Carrying cost estimation for various products, 108
Carrying cost increase, 13, 23, 80
Carrying cost reduction, 22, 82, 110, 215, 220, 222, 224, 228
Change triggered, 21–23
Change triggered by carrying cost, 219
Change triggered by demand, 222
Change triggered by ordering cost, 214
Change triggered by stock out cost, 220
Cost, 5, 7, 8, 10, 12, 13, 20–22, 30, 36, 46, 48, 51, 52, 55, 56, 58, 59, 61, 63, 67, 68, 72, 73, 76–78, 80, 82, 85, 89, 92, 97, 98, 101, 103, 106, 111, 113, 115, 117, 119, 126, 131, 134, 150, 151, 156, 169, 187, 189–191, 193, 196, 197, 200, 202, 203, 205, 206, 226, 229, 234
Cycle, 6, 16, 133, 135, 234
Cycle time, 6, 7, 14, 16, 131, 133–135, 139, 142, 145, 150, 154–157, 159, 161, 163, 165, 169, 174, 181, 183
Cycle time variation, 135

D
Decreased cycle time, 155
Decreased purchase cost, 120
Decrease in demand, 25, 30–32, 37–39, 66, 147, 148, 150, 161, 163, 210
Decrease in holding cost, 76, 77, 155, 157
Decrease in inventory carrying cost, 147, 149, 150, 161, 163
Decrease in ordering cost, 23, 151, 154, 160, 163, 165, 212
Decrease in stock out cost, 147, 149, 161, 167
Demand, 1, 3, 4, 9, 10, 20, 22, 25, 28, 30, 32, 37, 39, 41–43, 47, 52, 61, 62, 64, 65, 85, 87, 97, 105, 112, 133, 135, 137, 141, 159, 167, 176, 187, 198, 200, 214, 215, 219, 229, 233, 234
Demand estimation, 40, 43
Demand variation, 209
Downward variation in cycle time, 151, 154

E
Economic lot size, 11, 15, 18, 20
Estimation of ordering cost, 45, 47, 48, 69

F
FMCG, 40, 41, 67, 108, 232, 233
Food products, 67, 68, 109

H
Higher cycle time, 157
Higher demand, 25, 115, 137, 139, 155
Higher inventory holding cost, 137, 138
Higher ordering cost, 131, 139–141, 157, 160
Hi-tech, 40, 43, 67, 69, 108, 110, 234
Holding cost increase, 22, 73, 81, 85, 212

S. Sharma, *Inventory Parameters*, DOI 10.1007/978-981-10-4545-5

I

Inclusion of shortages, 54
Inclusion of stock out, 55, 64
Increased cycle time, 139, 141, 158, 161
Increased purchase cost, 122, 123
Increase in demand, 21, 25, 27, 28, 30, 35, 36, 58, 59, 61, 64, 83, 85, 105, 106, 137, 138, 151, 152, 154, 155, 159, 161, 163, 167, 176, 198, 209, 215
Increase in holding cost, 21, 73–75, 87, 89, 212
Increase in inventory holding cost, 73, 153, 154, 171, 181
Increase in ordering cost, 22, 23, 51, 53, 57, 92, 147, 150, 161, 162, 211
Increase in stock out cost, 151, 153, 154, 171, 183, 191, 198, 202, 206, 227
Interaction of holding cost with other parameters, 83
Interaction of inventory parameters, 20
Interaction of ordering cost and demand, 58
Interaction of parameters without stock out, 155, 161
Interaction of parameters with stock out, 161
Interaction of procurement cost with other parameters, 124
Interaction with the demand, 198
Interaction with the holding cost, 205
Interaction with the ordering cost, 201
Interaction with the parameters, 197
Inventory, 1, 3, 8, 9, 11, 17, 22, 32, 58, 71, 73, 80, 84, 109–111, 119, 126, 131, 133, 139, 143, 171
Inventory carrying cost, 8, 22, 72, 108
Inventory cycle, 13, 14, 133
Inventory cycle with stock out, 16
Inventory stock, 9, 32, 110

L

Lower cycle time, 171
Lower demand, 139, 140, 157, 160
Lower inventory holding cost, 139–141, 157, 158
Lower ordering cost, 131, 137–139, 155, 156

M

Management response, 21, 22, 61, 97, 100, 199, 209

O

Ordering, 7, 46, 49–52, 56, 67
Ordering cost, 6, 7, 11, 14, 20–22, 27, 45, 47, 51, 52, 57
Ordering cost decrease, 21, 23, 48, 212, 229
Ordering cost estimation for variety of products, 67
Ordering cost increase, 51, 52, 54, 58, 160, 211
Ordering cost variation, 211

P

Pharmaceutical, 40, 41, 67, 108, 109, 232, 233
Procurement, 2, 4, 9, 10, 21, 33, 39, 52, 58, 63, 71, 73, 75, 77, 105, 115, 135
Procurement cost, 5, 8, 111, 113, 117, 119, 122, 124, 127, 193
Procurement cost increase, 115, 116, 118, 122, 124, 127
Procurement cost reduction, 112–114, 120

R

Reduced cycle time, 135–137, 139, 146, 155, 156
Reduction in ordering cost, 21, 22, 48, 54, 61, 65, 95, 100, 101, 105, 156, 158, 167, 169, 171, 172, 174, 215
Response, 21–23, 97, 99, 214, 220, 222, 224, 226, 227, 229
Response with a combination of parameters, 97, 214

S

Shortage cost, 10, 188–190, 194, 197, 208–212
Shortage cost decrease, 193, 194
Shortage cost increase, 190
Significance of cycle time, 131
Similar lot size, 21, 22, 84, 85, 87, 89, 91, 95–97, 100, 102, 105, 124, 129, 197, 198, 200, 203–205, 207, 209, 210, 212–217, 219
Similar total cost, 21–23, 61, 65, 89, 91, 92, 94, 97, 99, 102, 197, 200–203, 205, 208, 209, 212, 214, 223–226, 230
Stock level, 9, 109
Stock out, 1, 10, 16, 17, 19, 20, 33, 39, 54, 65, 66, 83, 91, 97, 104, 121, 123, 146, 156, 160, 187–189, 193–195, 232, 234
Stock out cost, 10, 11, 22, 35, 188, 191, 194, 196, 201, 214, 215, 224, 233
Stock out cost estimation, 187
Stock out cost estimation for various products, 232
Stock out Inclusion, 119
Stock out level, 9, 10, 82, 122
Stock out situation, 11, 15, 16, 20, 22, 32–34, 37, 58, 112, 142, 233

T

Total annual cost, 14, 17, 111, 133

U
Upward variation in cycle time, 147, 150

V
Variation triggered by carrying cost, 226
Variation triggered by demand, 229
Variation triggered by ordering cost, 224
Variation triggered by stock out cost, 227

W
Without stock outs, 20, 21, 97, 156, 160
With stock outs, 22, 34–37, 58, 66, 82, 83, 86, 102, 108, 121–123, 161, 167, 169, 171, 174, 178

MIX
Papier aus verantwortungsvollen Quellen
Paper from responsible sources
FSC® C105338

If you have any concerns about our products,
you can contact us on
ProductSafety@springernature.com

In case Publisher is established outside the EU,
the EU authorized representative is:
Springer Nature Customer Service Center GmbH
Europaplatz 3, 69115 Heidelberg, Germany

Printed by Libri Plureos GmbH
in Hamburg, Germany